Mysterious Creatures

Terrifying First Hand Encounters With Bigfoot

(A Collection of Real Life Encounters With Bigfoot, Demons and Many More)

Charles Snyder

Published By **Darby Connor**

Charles Snyder

Mysterious Creatures: Terrifying First Hand Encounters With Bigfoot (A Collection of Real Life Encounters With Bigfoot, Demons and Many More)

ISBN   978-1-998927-73-9

No part of this guidebook shall be reproduced in any form without permission in writing from the publisher except in the case of brief quotations embodied in critical articles or reviews.

Legal & Disclaimer

The information contained in this book is not designed to replace or take the place of any form of medicine or professional medical advice. The information in this book has been provided for educational & entertainment purposes only.

The information contained in this book has been compiled from sources deemed reliable, and it is accurate to the best of the Author's knowledge; however, the Author cannot guarantee its accuracy and validity and cannot be held liable for any errors or omissions. Changes are periodically made to this book. You must consult your doctor or get professional medical advice before using any of the suggested remedies, techniques, or information in this book.

Table Of Contents

Chapter 1: What Cryptozoology Is and What It Is Not... 1

Chapter 2: Sightings of Pterodactyls and Other Strange Flying Creatures 9

Chapter 3: Mythological Creatures in the American West 33

Chapter 4: Lake and Sea Monsters......... 44

Chapter 5: Wild Men and Missing Links . 69

Chapter 6: ABCs and other OOPAs 95

Chapter 7: Too Strange For Classification 101

Chapter 8: Early Explorers Before Columbus... 116

Chapter 9: Aliens and UFOs.................. 140

Chapter 10: Strange Creatures............. 153

Chapter 11: Hauntings 172

Chapter 12: The Witch Who Got Away 178

Chapter 1: What Cryptozoology Is and What It Is Not

Cryptozoology is the technological know-how that investigates as however unconfirmed creatures, which investigators generally name "cryptids." The phrase is manufactured from the Greek stem "kryptos" (hidden) and suffix "zoology" (have a study of animals), as a cease result cryptozoology is the have a look at of hidden, as-but-undiscovered animals. It is critical to be precise in our definition due to the truth there's a exquisite deal of well-known false impression of what cryptozoology is and what cryptozoologists do.

Contrary to well-known opinion, cryptozoologists do now not have a have a look at the magical. They aren't ghost hunters or ufologists. While many cryptozoologists are interested by such phenomena, their very own work specializes in finding flesh and blood animals which have not been placed by using technological expertise. Visitors from different planets or from past the grave do not problem them. Some of the main researchers are themselves scientists, with advanced stages in zoology or biology, and but reject mainstream technological information's assumption that every one massive species have already been observed and catalogued. Other cryptozoologists aren't so careful, and the field is unluckily complete of a huge form of credulous, sloppy researchers.

Cryptozoology is likewise hampered with the resource of the fact that there were such masses of hoaxes. One of the most well-known come to be a weird beast that made headlines in 18th century France. In 1784, Courier de L'Europe cautioned on a weird animal captured in South America:

"This creature changed into located within the state of Santa Fe, Peru, within the province of Chili and in the lake of Fagua, located inside the lands of Prosper Voston. It emerged at some point of the night to consume the swine, bulls and cows of the location. Its duration is eleven toes; its face is more or lots less that of someone; its mouth is as large as its face. It has the horns of a bull and tooth inches prolonged. Its hair reaches to the floor. It has the ears of an ass, bat-like wings and tails, one flexible sufficient to seize prey, the opportunity completing in a dart which allows it kill. Its complete frame is included with scales.

"It grow to be netted and brought alive to the viceroy, who each day nourished it with a steer, a cow and three or 4 swine, to which it's miles pretty partial. The viceroy has despatched orders along the entire land course to offer for the dreams of this particular monster even as making it march by using manner of the use of degrees to the Gulf of Honduras, in which it will embark for Havana. From there to the Bermudas, to the Azores, and in 3 weeks it'll disembark at Cadiz. From Cadiz it's miles going to be taken via the use of short trips to the royal own family. It is hoped that a female can also be captured in order that the species will not die out in Europe. The species appears to be that of harpies, heretofore considered mythical."

An antique instance of a harpy

A Harpy, wings displayed.

The "harpy" by no means arrived in Cadiz or anywhere else, due to the fact the newspaper had definitely made the tale up. At that point, Paris became awash with newspapers, as many as hundred by way of the use of using some estimates, and that they competed fiercely for readers. One manner to sell papers turn out to be to make up outlandish testimonies, whether or now not approximately the affection lifestyles of public figures or weird guys and beasts placed inside the New World. The latter shape of tale have become the safer one. No one may take the paper to court, and as it took weeks to pass the sea, the ones stories have been tough to disprove. Despite going via the awakening of studying and technological expertise known as the Enlightenment, Europe although regarded out on a vast global with many smooth regions on the map. Who knew what might be hiding within the ones unexplored regions?

Cryptozoology additionally skips folktales that haven't any contemporary sightings and antique legends like the Pegasus that do not have any cutting-edge-day equivalents. While they may credit the ones memories with the possibility of fact, even the maximum credulous cryptozoologist doesn't anticipate he'll find out a dragon or the Medusa. And then there are the monsters no essential person became predicted to take into account in, made as a whole lot as entertain the folks around the campfire or to scare children into proper behavior. One of the maximum horrifying of these modified into an vintage tale from my non-public kingdom of Missouri—Raw Head and Bloody Bones.

This changed right right into a particularly nasty spirit that slunk spherical houses in the Ozarks in the 19th and early 20th centuries. The tale originated in England and become added to America via the early settlers, which allowed it to linger for awhile in the difficult hills and forested valleys of the Ozark Mountains in southern Missouri and northern Arkansas. It might also want to cover in the closet (of course!) or inside the drain pipe. If you've got were given been brave or stupid sufficient to peek inner at the proper time, you would likely capture a glimpse of a dark creature crouching there, looking back at you with its face dripping in blood.

Raw Head and Bloody Bones hated youngsters, and even as a little one changed into by myself within the house, it would turn the child right into a stain on the floor, possibly a drop of cooking oil or a spot of jam. When Mother and Father came domestic, they may scowl on the mess and wipe it easy, complaining all the at the identical time as approximately their messy offspring.

It ought to take a infant psychologist to make up some element more damaging to a growing mind than this story.

Chapter 2: Sightings of Pterodactyls and Other Strange Flying Creatures

When white, black, and Hispanic settlers moved into North America, they often discovered stories from Native Americans into their very private folklore. The Native Americans had been right here first, in the end, and whilst some brushed off their reminiscences as "community superstition," others believed the actual populace of the land knew some of its secrets and techniques.

One of the more enduring testimonies is that of the Thunderbird, a huge supernatural bird with wings so big it makes a thunderous noise while it flies. There had been many sightings of massive birds in the Old West over time, and it appears the settlers meshed European memories of dragons with the Thunderbird legend.

Newspaper hoaxes have often introduced extensive hobby to perceived creatures, and one regular tale within the Tombstone Epitaph on April 26, 1890, captured the readers' hobby within the Arizona Territory:

"FOUND IN THE DESERT

"A STRANGE WINGED MONSTER DISCOVERED AND KILLED ON THE HUACHUCA DESERT

"A winged monster, identical to a huge alligator with an incredibly elongated tail and a awesome pair of wings, turn out to be determined on the barren region some of the Whetstone and Huachuca mountains remaining Sunday by ranchers who have been returning home from the Huachucas. The creature come to be manifestly appreciably exhausted via an extended flight and at the same time as positioned, come to be capable of fly however a quick distance at a time. After the primary shock of wild amazement had passed, the two men, who have been on horseback and armed with Winchester rifles, regained sufficient braveness to pursue the monster and after an exciting chase of severa miles succeeded in getting near sufficient to open fireplace with their rifles and wounding it.

"The creature then have grow to be on the guys, however due to its exhausted scenario they have been capable of keep out of its manner and after a few nicely directed photos, the monster in part rolled over and remained immobile. The guys carefully approached, their horses snorting with terror, and placed that the creature have become dead. They then proceeded to make an exam and determined that it measured about 90- toes in length and the greatest diameter end up approximately fifty inches. The monster had best feet, those being placed a brief distance within the the front of in which the wings have been joined to the body.

"The head, as near as they might select, became about 8 ft prolonged, the jaws being thickly set with strong, sharp enamel. Its eyes were as large as a dinner plate and protruded approximately 1/2 way from the top. They had a few hassle in measuring the wings as they were in part folded under the frame, but in the end were given one straightened out sufficiently to get a size of seventy-8 toes, making the overall duration from tip to tip approximately a hundred and sixty feet. The wings have been composed of a thick and almost apparent membrane and had been without feathers or hair, as turn out to be the entire body. The pores and pores and pores and skin of the frame became relatively smooth and without problem penetrated via a bullet.

"The men lessen off a small portion of the top of 1 wing and took it home with them. Late very last night time time one in every of them arrived in this city for components and to make the critical arrangements to pores and skin the creature, at the same time as the quilt can be sent east for exam via manner of way of the eminent scientists of the day. The finder decrease once more early this morning located thru numerous distinguished men who will enterprise to deliver the uncommon creature to this town earlier than it's far mutilated."

No other article in this uncommon creature ever appeared within the Epitaph, although it's consistent to anticipate this type of find out might likely appeal to the interest of neighborhood scientists. By 1890, Arizona modified into losing its frontier revel in as more human beings settled there. A university had opened in Tucson, surely 70 miles away and a short journey through teach, five years before. One may also want to assume the professors there could have rushed to take a look at the creature within the occasion that they perception the object modified into real. It appears this tale modified into in the nature of a funny tale, as this sort of discover may have made it into the dominion and national press, no longer to mention medical journals. Also, the ranchers should were well conscious that the preserved cowl of the creature need to sell for a

immoderate fee.

Artist's rendering of a Thunderbird, thru Carnby

That article doesn't mention any pictures, but numerous images of intended Thunderbirds have emerged over the years. Some display a huge fowl tacked to the difficulty of a barn, while others display what seems more like the pterodactyl acquainted to paleontologists. There has been an outpouring of these on the Internet in contemporary years manner to the convenience of the usage of image software program program in which includes snap shots to modern vintage snap shots. One indicates a farmer defensive up a bit pterodactyl that seems like someone reduce open an umbrella and fixed a buzzard on the internal. Even pictures seemed to be fake, collectively with one displaying a group of Civil War soldiers proudly reputation beside a pterodactyl they've shot down (created for the Fox TV display Freaky Links), have generated enthusiastic speak on cryptozoology communicate boards.

This monster, which corresponds carefully to the Native American legend of the Thunderbird, ought to without troubles be unnoticed because of the truth the crafted from a few hack journalist who not first-rate had his tongue in his cheek, however a honest quantity of redeye whiskey as properly.

The reality that severa terrific newspapers in the Old West published their personal tales of Thunderbirds or dragons doesn't assist matters; it simplest proves that hoax journalism have end up a not unusual practice. Indeed, the Tombstone incident wasn't the first rate Thunderbird sighting. The Gridley Herald, a California newspaper, said in 1882 that men named Joseph Howard and Thomas Campbell were reducing wooden near Hurleton, California when they observed some thing flying above the treetops. It gave the impression of a crocodile, turn out to be

18 ft prolonged with a forty-foot wingspan, and had six wings and 12 toes. Howard claimed he fired at it along together along with his shotgun, and he instructed the newspaper, "It uttered a cry much like that of a calf and endure blended, but gave no sign of being inconvenienced or injured. In truth, while the shot struck, we heard the bullets rattle as although putting in competition to a thin piece of sheet iron." The editor opined, "This is the primary time we have got ever heard of this sort of creature as this, however our informants are dependable men, as a cease end result we cannot doubt their statements."

This wasn't the only flying monster supposedly setting round California. There were continual testimonies of a creature living in Lake Elizabeth, and rumors about this beast, which can every fly and stay underwater, commenced with the

Mexican settlers. The Lake Elizabeth monster became stated to be as large as a whale, but with massive, bat-like wings, and the place turn out to be avoided because it end up believed the creature ought to consume all the farm animals.

Finally, inside the Eighteen Nineteen Eighties, a courageous man named Don Felipe Rivera attempted to capture the creature a very good manner to sell it for $25,000 to a circus. He positioned the monster through manner of the lakeshore and fired at it alongside along along with his Colt .Forty four, however as with the opposite California monster, the bullets bounced off, growing a ringing sound as even though they had been hitting steel. Don Felipe Rivera suggested that he accrued his bullets after the creature moved off, and they have been as flat as little pancakes. He said the creature had the pinnacle of a bulldog, six legs, and have come to be about forty five feet extended.

In 1891, there has been a sighting of multiple flying "dragons" in Fresno. They had been 15 toes prolonged and had rows of razor sharp teeth they used to bite chickens in half of of. There become a comparable sighting in Utah in 1903. Even in recent times humans although claim to appearance pterodactyls or extraordinary strange flying creatures in the West. There modified right into a wave of pterodactyl sightings in Texas lower back within the Seventies, so in the event that they didn't skip extinct thousands and heaps of years within the past, in all likelihood a number of the flying dinosaurs are despite the fact that hiding out in the Lone Star State.

There have become additionally a wave of such sightings in Illinois and Missouri inside the overdue Nineteen Forties across the identical time as America's first wave of UFO sightings. While some Americans had been seeing metallic discs that might have been from every distinct planet, others were seeing big winged beasts. The first such sighting end up on April 10, 1948, even as three humans in Overland, Illinois, noticed what appeared like a chook the size of a small plane. Another eyewitnesses noticed a similar creature over Alton, Illinois, some weeks later. There had been moreover sightings spherical St. Louis, Missouri.

The Nineteen Seventies and 1980s noticed a terrific many flying creatures in American skies. Illinois had a wave of sightings in 1977, one of which could have led to tragedy. On July 25 of that 365 days at spherical eight:30pm, 10-year-vintage Marlon Lowe have become gambling with of his pals in his decrease returned outside when they noticed big birds circling overhead. They had a wingspan of approximately 10 toes (three meters), with hooked talons and beak, black feathers and a white ring across the neck.

One of those monsters dove for the men and grabbed Marlon as his buddies jumped to safety into the swimming pool. Marlon changed into grabbed across the shoulders and lifted two or 3 ft within the air and carried about forty feet in advance than the screams of his mother concerned the beast away. The boy come to be no longer considerably harm in spite of the reality that he became understandably rattled. Several one-of-a-kind sightings in Illinois that one year cautioned equal creatures.

The preceding twelve months in Texas, patrolmen Arturo Padilla and Homer Galvan noticed a large birdlike creature gliding thru the air. It turn out to be hard to see as it grow to be midnight, but the thing seemed reptilian with thin wings, no feathers, and an extended head. Three witnesses noticed the identical creature in Brownsville, Texas, in separate incidents. Soon sightings have been being suggested everywhere within the vicinity. Many stated it seemed like a huge bat and clearly no longer a bird, because it had no feathers. One schoolteacher suggested that the creature had a 12 foot (4 meter) wingspan and regarded like a pterosaur, a possibility we are in a position to speak later. Sightings hold, with numerous eyewitnesses seeing the creature within the skies round San Antonio, Texas, in 2007.

These are the form of critiques that intrigue cryptozoologists and this is the shape of evidence we are able to be looking at on this e-book. While legends and folklore may be used as helping evidence to suggest that massive flying creatures had been in our skies for a long time, the similarity in current sightings over several a few years and a large geographic vicinity suggests to cryptozoologists that there is probably some foundation of fact to the antique recollections.

Several opportunities have furnished themselves. The maximum not unusual is that the eyewitnesses are not mistaken in what they located, but they may be seeing a referred to animal, in particular the Andean condor, which has a wingspan of up to 10 feet (3 meters). This fowl, the area's biggest, appears similar to the chook that grabbed more younger Marlon Lowe. It isn't always past the region of opportunity that a condor strayed from its traditional habitat in South America and made it all the manner as tons because the USA. While the diverse sightings claim the creature modified into an lousy lot huge, it's miles hard to decide the gap and length of a flying object, plus the shock of seeing this kind of aspect may additionally have made the eyewitnesses unconsciously exaggerate what they observed. Others item that an Andean condor isn't robust sufficient to select out up a little one, but it's doubtful if any experiments have ever been made to check this! Besides, it regarded that the bird had problem lifting terrible little Marlon, flying almost degree and coffee to the floor for some distance without being able to ascend. A fowl of prey normally ascends right away after grabbing its lunch.

So even as some Thunderbird sightings may be condors which have strayed a ways from their ordinary habitat, many eyewitnesses say that the creatures are reptilian, without any of the competencies of a condor or any fowl. A extra debatable rationalization held with the useful resource of using a few cryptozoologists is that the creatures may be pterosaurs, a shape of flying dinosaur that first seemed round 228 million years inside the past and supposedly went extinct round sixty five million years within the past. This lengthy-lived beauty of lizards have been the primary diagnosed vertebrates to have evolved winged flight. There were many extraordinary species over the good sized duration of time they dominated the skies, from the Nemicolopterus crypticus ("Hidden flying wooded vicinity dweller") that had a wingspan of pleasant 10 inches (25 cm), to the formidable Quetzalcoatlus northropi (named after the Mesoamerican feather serpent god and an aviation

professional) that had a wingspan of 36 feet (eleven meters).

A sordes, a pterosaur

The United States isn't the high-quality u.S. Of the united states to be blessed with pterosaur sightings. In his e-book In Witch-positive Africa, written in 1923, explorer Frank Melland writes approximately the customs of Rhodesia, what's now Zimbabwe. The humans there dreaded the kongamato ("overwhelmer of boats"), a flying monster that could swoop down and turn over boats crossing the river. Anyone the use of a deliver needed to prepare a completely precise attraction to shield themselves.

Melland wrote:

"After hearing about the attraction that's applied in opposition to the kongamato, and the powers attributed to it, I asked 'What is the kongamato?' The answer emerge as, 'A hen.' 'What form of bird?' 'Oh, nicely it isn't a chicken definitely; it's far greater like a lizard with membranous wings like a bat.' I write this down verbatim due to the fact I end up not within the least thinking about, nor looking for for, facts as to any prehistoric animals or reptiles. It came on me quite all at once…

"…Further enquiries disclosed the 'statistics' that the wing-spread grow to be from 4 to seven ft at some stage in, and that the overall shade changed into crimson. It emerge as believed to don't have any feathers but only pores and skin on its frame, and changed into believed to have enamel in its beak: those very last elements no character can also moreover need to ensure of, as no man or woman ever noticed a kongamato close and lived to inform the tale. I despatched for 2 books which I had at my residence, containing snap shots of pterodactyls, and every nearby present without delay and unhesitatingly picked it out and identified it as a kongamato. Among the natives who did so grow to be a headman (Kanyinga) from the Jiundu the us, in which the kongamato is supposed to be energetic, and who's a instead wild and pretty unsophisticated local...

"...The evidence for the pterodactyl is that the natives can describe it so appropriately, unprompted, and that every one of them agree approximately it. There is poor resource additionally inside the fact that they stated they couldn't pick out every one of a kind of the prehistoric monsters which I confirmed them...

"...The cutting-edge regarded (?) case of its sports is stated to have been in 1911, at the same time as men and girls were killed with the aid of way of a kongamato in the Mutanda River close to to Lufumatunga village.

"…I even have stated the Jiundu swamp as one of the reputed haunts of the kongamato, and I even have to mention that the area itself is the very kind of place in which this form of reptile may want to probably exist, if it's miles feasible everywhere. Some fifty square miles of swamp, fashioned by means of using an inland delta, as demonstrated within the accompanying cool animated film map: the Jiundu River spreading out into innumerable channels, and — after receiving numerous tributaries, reuniting further down right right into a unmarried circulation of crystal-easy water. Much, however, that flows into the swamp does not pop out, however vanishes beneath ground, as many rivers and streams do in the ones additives. The entire of the swamp is covered with dense flora: large timber that develop to a excellent top, tangled undergrowth with matted creepers and super ferns. The soil is moist loam and decaying plant life, the number one channels and lesser rivulets being harking back to the peat hags on a highland moor. It is a replica on a big scale of the groves of swamp wood that one sees via maximum streams (jitu) but which I, in my view, have now not met with in such duration except with the useful aid of the Semliki mouth in western Uganda. The complete of the u . S . A . Is of a limestone formation, and out of doors the swamps, even among the stretches of timber, outcrops are sufficient. In one vicinity sincerely to the west of the region there can be a massive hollow similar to a crater. Nowhere else on a excessive, properly-worn-out floor have I visible this sort of morass nor must one conjure up a greater ideal photograph of a haunted wooded vicinity.

"If there be a kongamato, that is surely a superb home for it."

Melland turn out to be no substantial-eyed traveller. A graduate of Oxford University, he changed right into a fellow of the Royal Geographical Society and the Zoological Society. Elsewhere in his e book, he dismisses the idea of another monster, this one a brontosaurus type creature, which he said couldn't have remained hidden from explorers. A pterodactyl (a kind of pterosaur), he reasoned, is probably able to cover extra with out problems.

But should a creature that supposedly died out 65 million years in the beyond in spite of the fact that be spherical these days? Science has confirmed that severa animals have finished simply that. The horseshoe crab, as an example, has remained pretty unchanged for 450 million years. The coelacanth, a form of fish, has been spherical for 410 million years and modified into idea to have gone extinct within the Late Cretaceous until it modified into rediscovered inside the twentieth century. Compared to those, the bullhead shark is a relative teen, having been spherical for "handiest" 183 million years. There are many other examples of such "residing fossils" that have advanced slowly over a giant span of time and still carefully resemble their fossil ancestors.

Chapter 3: Mythological Creatures in the American West

The Thunderbird wasn't the most effective monster to seem in the Wild West. According to the lore, there has been no shortage of unknown creatures, some horrifying, a few bizarre, and a few downright ridiculous.

In the ridiculous class is the fur-bearing trout, a tale that got here from the mining camps of Colorado. There changed into one camp in which an unusually high variety of the miners have been bald. Whether the miners had been antique or truly unlucky the story doesn't say. A hair tonic salesman headed to the camp inside the destiny that permits you to "mine the miners", due to the fact the vintage announcing is going. During a gold rush (or perhaps a lead rush), mining the miners have become the brilliant way to strike it rich, as mining camps sprung out of now not some thing and everything modified into in short deliver. Anyone with a wagonload of canned goods, tools, and liquor must make coins with extra fact than even the most diligent digger, so this hair tonic salesman figured this colourful-scalped mining camp can be his Mother Lode. He have emerge as out to be smarter than he emerge as lucky, however, because of the truth as he modified into crossing the Arkansas River he by chance dropped his product within the water. The bottles broke and his hair tonic stuffed the river. Soon neighborhood fishermen decided that the trout had grown hair, so in region of taking off a hook and bait, they installation barber poles with the aid of manner of the element of the river. The fur-bearing trout, desperate to put off their hair and bypass back to their natural scaly baldness, leapt out of the water and the fishermen collected them.

Some say the bald miners desired to get a bit of that hair tonic and rubbed the fur-bearing trout on their hairless pates inside the hopes of a treatment. One penniless miner named Shoeless Magee, whose ft had been as naked as his head, thru accident dropped a fish on his feet and ended up with hair on his ft further to his head. He later hit the Mother Lode and retired in San Francisco. He attributed his trade of fortune to the brand new hair the fur-bearing trout gave him and for the relaxation of his days is probably seen on foot the streets of San Francisco barefoot.

Then there's the Jackalope. This critter has the frame of a jackrabbit and the antlers of a pronghorn antelope, for that reason the decision. Most are the size of everyday jackrabbits, but again in the antique days a few grew so massive that cowboys might also moreover need to located bridles on them and hop across the prairie. There are postcard pix supposedly proving it, and postcard images can also want to by no means lie. Of route, catching and taming the not unusual Jackalope isn't any easy challenge. Jackalopes are timid creatures and fool their pursuers thru imitating human speech, pronouncing things like "He's run inside the returned of that tree over yonder!" on the identical time as in fact the critter had run behind a rock inside the contrary course. Some say the extraordinary way to entice him out is with whiskey.

Naysayers and curmudgeons declare that the Jackalope modified into genuinely invented through a hunter and amateur taxidermist named Douglas Herrick (1920–2003) from Douglas, Wyoming, who in 1932 killed a jackrabbit and decided to maintain it with a similarly touch: multiple antelope horns. He offered it to a community resort and shortly his

Jackalopes have been a huge hit. In time, the Herrick own family turned into selling Jackalopes thru the thousand, or even the city of Douglas were given in on the sport through selling Jackalope searching licenses.

An example of a horned rabbit

The Ozarks in northern Arkansas and southern Missouri have lengthy been a hideout for outlaws, Confederate guerrillas, and weird legends. The thinly populated location has thick underbrush, tough hills, deep valleys, and big swamps, so it's occasionally sudden that locals tell many a weird story approximately what lurks in the ones hills. One of the strangest is about the Boggy Creek Monster. This is the nearby Bigfoot, a seven-foot-tall hairy beast with a robust, musky smell. It's determined near the area of the town of Fouke and has regularly been sighted throughout the creek that gives the monster its call.

Tales of the Boggy Creek Monster move lower back to the 1840s, and it became recognized for killing livestock and puppies, maximum probable for meals. It come to be commonly feared and lots of locals averted the creek. Sightings endured for a century, so likely there's a whole own family of Boggy Creek Monsters down there. A farmer who observed it in 1973 stated it have turn out to be best four toes tall, but maximum sightings say it's man-sized or big, so this will had been a infant. One female who observed it in 1971 said it had extended, darkish hair and walked hunched over, swinging its hands like a monkey.

There modified into a wave of sightings in 1996, however the ones have died down in modern years. Sadly no photos exist, so any proper believer wishes to be content fabric fabric with a 1972 movie referred to as The Legend of Boggy Creek. The movie made a lovely sum of money and extra Boggy Creek films had been made.

While the Boggy Creek Monster in no manner attacked human beings, one in every of Arkansas' one among a kind legendary beasts isn't always so outstanding. According to legend, unusual being lurks beneath the muddy waters of the White River near the Ozarks metropolis of Newport, 80 four miles northeast of Little Rock. Known due to the fact the White River Monster, locals affectionately talk to it as Whitey. Most sightings describe it a huge serpent about 30 toes prolonged with spines along its another time. Three fishermen who noticed it in 1966 stated it had the pinnacle of a monkey and arm-like flippers, and plenty of have remarked at the high-quality sounds it makes, described as a bellowing or a mixture of a cow's moo and the neigh of a horse. The monster became recognized to the Quapaw Indians and as soon as overturned one in every of their canoes. During the Civil War, at the same time as Arkansas have become part of the Confederacy, Whitey proved himself to be a Yankee via the use of overturning a Confederate gunboat.

Modern sightings, documented through community newspapers, began out in 1915, and there had been sightings on and stale ever thinking about the reality that. There have been approximately 100 sightings in 1937, and one guy who claimed he observed it turned into Bramlett Bateman, who owned a plantation with the useful resource of the river. He even opened up a viewing place, charging humans 1 / four to sit down down and stare out at the river. Being the hospitable kind, Bateman saved them refreshed with food and drink, for a rate of direction.

In the early 1970s there has been every other wave of sightings and the primary picture of Whitey, a blurry photograph showing what seems like an alligator with outsized eyes swimming alongside the river. In 1971, a few three-toed tracks measuring fourteen inches lengthy have been discovered alongside the muddy riverbank together with hundreds of overwhelmed underbrush as though a big animal had walked via it. Even a few small wooden had been damaged, so the critter need to had been quite sturdy.

In 1973, the Arkansas State Legislature created the White River Monster Refuge to shield the creature. It lies next to Jacksonport State Park. People can't "molest, kill, trample, or damage the White River Monster at the same time as he is inside the retreat."

In the wilderness Southwest, the Mexican community has handed down recollections about the Chupacabra, which translates to "goat sucker". This small humanoid creature has been part of Mexican way of life for hundreds of years and may be based mostly on Native American ideals. The earliest file comes from 1540 and describes them as small men with dark scales. Each carried a torch and a spear and they will carry together in huge numbers to assault settlements. Since then, the Chupacabra has been described in lots of techniques steady with the style of the time. In present day times he looks as if a gray alien with fins on the again of its head and backbone. As their name shows, they suck the blood out of goats and different farm animals.

Chupacabras became today's within the 1990s, with sightings all through Latin America, mainly Mexico, in addition to the Southwest. In fact, Chupacabras regarded to conform with the Mexican network anyplace they went, and there have been Chupacabra sightings in Mexican neighborhoods as a long way north because the Great Lakes area. Chupacabra t-shirts have become all of the rage and there was even a Chupacabra dance in Mexico.

Chapter 4: Lake and Sea Monsters

Without query, one of the maximum famous cryptids is the Loch Ness Monster. This glacial lake in Scotland has been generating regular sightings for the motive that early Middle Ages. The 7th century manuscript, Life of St. Columba, recounts the adventures of an Irish monk in the place as he transformed the pagan Picts and finished miracles, and this early description is nicely genuinely well worth quoting in whole:

"When the blessed man grow to be staying for some days within the province of the Picts, he observed it crucial to cross the River Ness; and, at the same time as he got here to the monetary group thereof, he sees a number of the inhabitants burying a bad unfortunate little fellow, whom, as individuals who were burying him themselves said, some water monster had a piece in advance than snatched at as he changed into swimming, and bitten with a most savage chew, and whose hapless corpse some men who got here in a boat to provide help, despite the fact that too late, stuck maintain of via the use of putting out hooks. The blessed man but, on paying attention to this, directs that a few one among his partners shall swim out and produce to him the coble [a type of small fishing boat] that is on the opposite financial institution, crusing it at some point of.

"On listening to this course of the holy and famous guy, Lugne Mocumin, obeying at once, throws all his clothes except his below-garment, and casts himself into the water.

"Now the monster, which before emerge as no longer so much satiated as made keen for prey, modified into mendacity concealed inside the bottom of the river;

but perceiving that the water above become disturbed through him who modified into crossing, unexpectedly emerged, and, swimming to the person as he end up crossing within the middle of the flow into, rushed up with a wonderful roar and open mouth.

"Then the blessed man appeared on, even as all who have been there, as well the heathen as even the brethren, have been with very high-quality terror; and, along alongside along with his holy hand raised on excessive, he shaped the saving sign of the bypass within the empty air, invoked the Name of God, and commanded the fierce monster, saying, 'Think no longer to transport in addition, nor contact thou the character. Quick! Go lower lower back!'

"Then the beast, on listening to this voice of the Saint, turn out to be terrified and fled backward more suddenly than he got here, as though dragged with the useful resource of cords, no matter the fact that before it had come so near to Lugne as he swam, that there was not more than the duration of one punt-pole the various person and the beast. Then the brethren, because of the fact the beast had long long past away, and that their comrade Lugne became lower returned to them steady and sound within the boat, glorified God in the blessed man, considerably marveling. Moreover additionally the barbarous heathens who have been there present, constrained by the greatness of that miracle, which they themselves had visible, magnified the God of the Christians."

The careful reader may be aware that this encounter occurred on the River Ness, one of the rivers that feeds the loch, and now not the loch itself. While many skeptics have used this reality to dismiss the tale of

St. Columba, in reality there have been numerous Nessie sightings on this river.

Between the Middle Ages and the Modern Age there had been few sightings besides for a smattering of money owed in the 19th century. A wave of sightings in the Thirties delivered global hobby to the loch and the beast has made limitless appearances ever at the same time as you endure in mind that.

The first image claiming to be of the creature become taken in 1933, and hundreds greater have been taken for the reason that then. Most display splendid a vague shape inside the water, looking like one or extra dark humps rising from the floor. A few show an extended neck with a highly small head. The maximum well-known and iconic photograph, the so-known as Surgeon's Photograph taken in 1934, become tested a faux while one of the people who helped take it made a deathbed confession. Many remarkable, comparable, photos have not all started to be debunked.

The important idea to provide an purpose for this phenomenon is that Nessie is a plesiosaur, a type of marine dinosaur that supposedly died out sixty 5 million years inside the beyond. The elegant description of Nessie, with its long neck and small head, its humped lower returned, massive fins, and lengthy tail, tally nicely with fossils of plesiosaurs. Several of the higher

pics seem to reveal this kind of creature. How those animals survived for goodbye is a thriller. During a amazing deal of the time that plesiosaurs are appeared to have lived, this vicinity of Scotland become below water. A plesiosaur fossil courting lower back one hundred and fifty million years emerge as placed at Loch Ness in 2003, but the loch itself wasn't carved out till the final Ice Age 12,000 years within the beyond. How the creature were given into the lake is every extraordinary thriller.

The famous Loch Ness monster hoax photo of 1934

Accounts of the monster variety depending on how simply and from what distance the beast have become noticed, however they generally agree on the primary facts. One very easy sighting on July 22, 1933 via manner of the use of George Spicer and his companion describes how they have been the usage of along the lakeside avenue (which had great been constructed earlier that yr, bringing greater human beings to the previously remoted loch) and observed the creature skip their path and slip into the water. Its frame was about 4 feet (1.2

m) excessive and 25 toes (7.6 m) prolonged. Much of this era was taken up with the beneficial useful resource of the slim neck, which Spicer defined as being a bit thicker than an elephant's trunk. It grow to be so long as the street was enormous, making it approximately 12 feet (4 m) lengthy. It modified into sinuous and undulating because the creature moved. A dip in the street stored the Spicers from seeing the lowest of its frame, in order that they couldn't say if the creature had fins or toes.

That identical yr, veterinary student Arthur Grant claimed he observed it while on his motorcycle late one night time. While he didn't get a remarkable view due to the darkness, he defined the creature as looking like a circulate amongst a plesiosaur and a seal. Grant's testimony is substantial due to the truth he had zoological know-how.

10 years later, in the course of World War II, every extraordinary professional observer claimed he observed Nessie. This changed into C. B. Farrel of the Royal Observer Corps, a civil protection unit that stored a watch out for Luftwaffe planes. He saw a creature that emerge as 20-30 toes (6-9 meters) prolonged with fins and a neck that caught out approximately 5 feet (1.Five meters) from the water. Farrel took a large risk in making his file, because of the reality this have grow to be the peak of the Blitz and his advanced officials have been understandably on facet and could have had little patience for such things as lake monsters.

Since the early wave of sightings within the Thirties and '40s, there was a normal amount of stories, pix, and films ever for the reason that. What is top notch is how steady they'll be about the general form and period of the creature. There have additionally been sonar contacts with the aid of numerous researchers who have detected big devices shifting about deep in the water. Recently, but, there was a few problem that the Loch Ness Monster can be going extinct. Sightings had been dropping off seriously inside the first decade of the 21st century. A sonar sweep in 2003 did not come across any underwater presence.

Luckily for the monster and the neighborhood financial gadget, 2015 became a bumper year for Nessie sightings, with five formally diagnosed sightings. Two of the sightings covered photographic proof and were seen with the aid of the usage of multiple witnesses. The cryptozoologists, who make it their organisation to report sightings around the Scottish loch for the Official Loch Ness Monster Sightings Register, endure in thoughts a sighting actual if it defies easy rationalization.

Unlike many cryptozoologists, some of the greater first rate Nessie professionals are scientifically minded skeptics and push aside many sightings as waves or logs. With such lots of tourists coming to Loch Ness each twelve months, bringing an predicted £30 million ($forty three million) to the vicinity every 365 days, human beings are going to appearance what they want to appearance.

The first sighting of 2015 befell on April 22, at the equal time as a pair touring the ancient Urquhart Castle at the beaches of the loch saw what they concept gave the impression of a big dolphin emerge from the water, best to fast slip below the ground. This became repeated 4 to six instances earlier than the creature in the end disappeared. Another person who didn't apprehend the couple additionally observed the creature.

Just a few days in a while April 25, Dee Bruce and Les Stuart were driving alongside the shore road when they noticed Nessie come approximately 3 feet out of the water close to the north quit of the loch earlier than speedy disappearing yet again.

On July 1, Crystal Ardito become the use of in a boat at the loch whilst she saw a weird form in the water and photographed it. It turn out to be visible for just a few seconds and she or he or he or he could not truely see what it modified into. When she later examined the image, she zoomed in and observed a grayish object poking out of the water. It modified into distant and consequently quite pixilated, but it appears to be a grey mass atop some other gray mass. True believers might say that it become Nessie's neck and hump shifting far from the photographer, however it is able to in reality be pretty a good buy something.

On August thirteen, Mr. & Mrs. Bates took numerous pics of a terrific undulating mass shifting across the water simply off from the Loch Ness Holiday Park. It end up visible for about 5 mins and five one of a kind humans furthermore noticed it. It did no longer seem like a wave, which wouldn't have lasted extended except, but due to the reality no longer a few element ever surfaced, it's far tough to say precisely what it changed into.

All of those sightings happened in daytime in clear conditions, and Gary Campbell, the keeper of the take a look at in, and Steve Feltham, who has been keeping a almost each day watch over the loch for 25 years, say that at the same time as they used to consider within the plesiosaur idea, they now suppose the monster is in truth a big catfish or eel, or to be greater precise a small populace of such fish that have been delivered into the loch in Victorian times.

Loch Ness isn't the handiest frame of water stated to harbor a monster. Of path there are the ocean monsters that the early sailors claim to have encountered, however they've got all however disappeared, whether from a lack of creativeness amongst contemporary sailors or all the pollution we've been pumping into the area's oceans.

One of the more enduring sea monsters is positioned off the coast of Devon and Cornwall in southwestern England, in particular around Cornwall's Falmouth Bay. Descriptions range, with many antique sailors saying it seems like a massive octopus or a conventional sea serpent similar to an oversized snake. Others say it seems more like Nessie, suggesting it might be a surviving plesiosaur. A few pics of the creature exist that certainly seem like a plesiosaur.

The creature come to be first cited on May 6, 1876, in a excellent article within the Royal Cornwall Gazette:

"Portscatho. THE SEA SERPENT!—Our correspondent, Mr. Bosisto, writes:—'The sea-serpent changed into caught alive in Gerrans Bay at the twenty sixth ult [the 26th of last month]! Two of our fishermen have been afloat overhauling their crab-pots, about four hundred or 500 yards from the shore, when they found a serpent coiled approximately the floating cork-mark of the crab-pot below. Upon their close to method, it lifted its head and confirmed signs and signs and signs of defiance, upon which they struck it forcibly with an oar, which up to now disabled it as to allow them to maintain with their artwork, and then they located the serpent floating approximately close to the boat. They pursued and captured it, bringing it ashore however alive for exhibition, quick and then it have become killed at the rocks and most inconsiderately strong all over again into the ocean.

"Why changed into now not the extraordinary creature, for which so many people have been searching, preserved and exhibited? It may have brought popularity to Portscatho and riches to its captors. 'There is a tide inside the affairs of fellows which, taken on the flood, consequences in fortune.' The Portscatho fishermen have left out their tide."

The serpent changed into visible several times after that. In one sighting on July 5, 1912, Captain Ruser of the German supply Kaiserin Victoria and his institution observed it off the coast of Devon. They stated it to be 20 ft (6 m) prolonged and 18 inches (46 cm) thick, with a blue-gray coloring and a white underbelly. The sailors said that it have come to be thrashing violently within the water, making its entire frame seen.

Sightings endured, with some reporting that it had a "horselike" or "beaked" head, or scaled legs. Others noted stubby horns. Colors numerous from black to inexperienced. There had been other variations in description as properly, dividing cryptozoologists as to whether or not or not the ones have been unreliable opinions of not some thing, or legitimate evaluations of numerous specific kinds of creatures.

In 1933, the rotting frame of the ocean monster washed ashore in Cornwall. It was so decayed that it end up difficult to say whether or now not or now not it become surely the mutilated remains of a few extra not unusual sea creature or some factor more sinister. Sailors and nearby scientists had been no longer capable of understand it.

Many of the greater cutting-edge-day reports describe some issue much like Nessie. It's thrilling that the creature converted from a conventional snakelike sea serpent into the extra iconic lake monster with its big frame, flippers, prolonged neck and small head. Skeptics say that is the crafted from notion, in which humans see some thing normal within the water and task their assumptions onto what they've seen. This concept is weakened thru the truth that in a wave of sightings in 1976, the various eyewitnesses described the creature as a big snake, in preference to a plesiosaur type creature.

While the Cornish sea serpent is one of the closing of a dying breed, inland lakes are no matter the fact that high breeding grounds for mysterious existence.

In the metropolis of Loughborough, in Leicestershire, England, a neighborhood lake have come to be the center of expert state of affairs in early 2010 whilst some of witnesses reported seeing ducks sucked into the water and never resurfacing. The chairman of the Charnwood Wildlife Protection Group showed that the lake's duck population had really reduced. The close by council warned people to hold their children and small animals a ways from the lake. The offender has been dubbed the "Lough Ness Monster," that is a bit unoriginal however truly greater charitable than "Lougie."

America might not be the residence of Loch Ness, but there are supposedly all varieties of lake monsters. It seems there isn't a lake, pond, reservoir, or outsized puddle that doesn't have its private monster skulking about in it. Cowboys used to say that if someone left an open barrel out inside the rain earlier than breakfast, there is probably a lake monster in it with the beneficial aid of suppertime.

One of the extra well-known American lake monsters is the Bear Lake Monster in Utah, and memories of this ninety-foot brown snake pass again almost two hundred years. It has a skinny head that's nearly totally one massive mouth, capable of swallowing a person in a unmarried gulp. It's been identified to eat swimmers however every now and then, if it's feeling playful, it's going to sneak up and blow a spout of water at them. It additionally has little legs along its sinuous body that it is able to use to scuttle alongside the land. One story claimed that a courageous hunter tried to shoot it together along with his rifle however neglected with every shot, as if the creature become magically deflecting the bullets. The hunter then did the smart difficulty with the useful resource of losing his gun and hightailing it to the following county. The Bear Lake Monster ate his rifle and sunk backpedal underneath the floor. It hasn't been seen for decades, so possibly it died of indigestion.

An even stranger lake monster is the Alkali Lake Monster. Alkali Lake, now referred to as Walgren Lake, is determined in a volcanic region of Nebraska. The lake is high-quality 50 acres in duration and is famous among fishermen for its bullhead and largemouth bass, however it's a stranger fish that has earned notoriety. The Native Americans had been properly aware about some element unusual in the lake, some element that appeared like a forty-foot gator with a gray pores and pores and pores and skin and a big horn on its snout. Not even the bravest courageous may paddle out into the lake's waters for fear of that horrible horn ripping thru the bottom of his canoe. It can be lunchtime then, and not for the nearby.

Like lots of America's lake monsters, this one has survived into the modern technology. A sighting stated within the Omaha World Herald on July 24, 1923, through way of a Mr. J.A. Johnson stated that he and partners observed the monster swimming thru the lake. It must have determined them too, because it let out an ear-splitting roar and disappeared below the ground. Johnson stated the creature have become extensively identified to the humans of the place: "I must call forty one in all a kind people who have moreover visible the brute."

Asia additionally has numerous lake monsters. In Lake Tianchi, within the mountainous border among China and North Korea, there's an entire populace of monsters, with one eyewitness account claiming up to 20 creatures. The sightings date all of the manner once more to 1903 and describe a creature that looks like a big buffalo that rushes at those who dare visit the lakeshore. While Lake Tianchi is in a much off and politically sensitive location, Japan's Lake Ikeda is cautiously used and without issues on hand. For the past thirty years there had been numerous sightings of a plesiosaur kind creature that has truely been dubbed "Issie." Even the U.S. Military had a run-in with the creature in one of the earliest encounters. Back in 1961, a navy jet crashed into the lake and whilst American divers used sonar to look for the wreckage underneath the floor, they caught sight of a large creature transferring within the depths. Of direction American servicemen aren't afraid of something, so divers went all the manner down to look for the wreckage except, and nearly had been given attacked via the beast. Apparently Central

Command decided it wasn't a chance to country wide safety and didn't bomb the lake.

While those are the most well-known examples, now not all lake monsters are restrained to the northern hemisphere. There are numerous in South America and Africa as properly. In Nahuel Huapi Lake, in Patagonia, Argentina, there's Nahuelito, a Nessie-like being that has been sighted for the past few a long term and has been really photographed four instances. Over in relevant Africa, the Chipekwe or Emela-ntouka terrorizes villagers near shallow stretches of water. It can normally be discovered in shallow swamps, lakes, and rivers and looks like some sort of mutated cross amongst a massive lizard and a rhinoceros. The pygmies call it the "killer of elephants," after its desired prey, which gives an concept of the way fierce the creature is. During colonial times numerous expeditions of large sport hunters went after it, but without an excellent fortune.

Cryptozoologists talk of "monster range," the excessive areas in each the northern and southern hemispheres in which lake monsters may be located. This corresponds to latitudes wherein the isotherm (midrange temperature) for freshwater lakes is 50º F (10º Celsius). They trust that that is the first-rate ecozone for such monsters, and feature charted the evaluations of lake monsters to expose that the awesome majority of them fall internal the ones bands. It's an exciting notion, and provides a bit of scientific rigor to an otherwise fuzzy region of research.

Chapter 5: Wild Men and Missing Links

For a whole lot of human history, the great majority of land modified into carefully inhabited. In clearly all cultures there grew up stories of what lived inside the ones empty, unexplored areas. One of the maximum continual is that there are so-called "wild men," people or humanlike beings that live in the wasteland and exceptional once in a while be part of up with civilized humans. Sometimes they'll be opposed, however greater regularly they shy away from touch. The most famous of those in modern instances is the Sasquatch or Bigfoot of North America and the Yeti of the Himalayan Mountains in Asia.

The Sasquatch originated as a Native American legend, even though inside the ones testimonies he come to be certainly a huge human with long hair who lived out within the wild, not some simian creature with hair at some stage in its frame. While maximum sightings center throughout the Pacific Northwest, the creature has been seen in pretty lots each u.S. Within the U.S. And throughout all but the northern Arctic provinces of Canada.

Descriptions of the Sasquatch or Bigfoot usually say it's about 7-10 feet (2-3 meters) tall and guarded with black or darkish brown hair, every now and then with a reddish tint. It's significantly muscular, walks on two legs like someone but with a loping stride similar to an ape's. It is generally seen on my own however every now and then whole families, which include youngsters, have been stated. The head is an unusual characteristic, with a low brow, a outstanding forehead ridge, and a rounded or crested pate. Sometimes the eyes are defined as being very massive. The feet, of course, are quite big too, with footprints being measured as a bargain as 24 inches (60 cm) lengthy and 8 inches (20 cm) in width.

Sightings with the aid of using way of European settlers are as antique as European settlement itself, and progressed because the numbers of white guys increased in the Americas. In the twentieth and twenty-first centuries, there were masses of documented sightings and masses of footprints recorded, masses of them photographed or preserved with plaster casts.

Like with maximum wild man sightings, the creature usually actions away as quickly as it is observed, however this isn't continuously the case as a few miners in Washington decided out to their remorse in 1924. The guys had been staying at an isolated cabin within the Mount Saint

Helens Range about 75 miles north of Portland. While out surveying in the end, they noticed a big apelike creature looking them through the underbrush. Fred Beck, one of the miners who emerge as there that day, stated taken into consideration one in each of his companions grabbed a rifle and shot at it. He seemingly disregarded the creature, which moved fast off, on the side of severa other creatures who abruptly burst from the foliage to make their get away. Beck himself fired at certainly one of them, hitting it and knocking it off a cliff. A are seeking out observed no hint of the creature.

That night time, as the men bedded down of their windowless log cabin, they heard a horrible noise out of doors. There turn out to be a violent pounding at the door and partitions as if some massive beasts were hitting it with their fists. Then they heard

72

what seemed like large rocks being thrown at the cabin. None of the miners had the center to open the door to look what end up taking vicinity, as an alternative they clutched their rifles and stayed placed, praying it'd rapid be over.

The attack lasted numerous the night earlier than in the long run truly fizzling out. That morning there were big footprints anywhere within the cabin. The men didn't live to investigate, but, and beat a hasty retreat returned to civilization.

Since then, numerous big footprints had been photographed or preserved with plaster casts, and there had been limitless sightings at some point of North America. Most of the pictures are vague, and will without issues be neglected as shadows or tree stumps.

Less easy to disregard is what has end up called the Patterson-Gimlin Film. On October 20. 1967, Roger Patterson and Bob Gimlin had been the usage of their horses alongside the Klamath River in the northwestern corner of California after they came throughout a Bigfoot at the back of a big fallen tree. Gimlin related that he have end up so shocked that he did nothing, while Patterson had been given off his horse, grabbed his 16mm film digital digital camera, and commenced out filming.

The quit end result became the most arguable video of a cryptid ever recorded. The movie runs for a piece an awful lot tons much less than a minute and certainly suggests a large, stocky apelike being strolling some distance from the virtual camera with a loping stride. In one chilling 2nd, it appears over its shoulder on the digital camera and the face can sincerely be visible. That face looks like a mixture of human and ape capabilities. The creature seems to be a girl due to the fact the breasts are pretty terrific.

Patterson stated he decided the Bigfoot for some time, taking the photos which can be visible inside the film. When Gimlin didn't follow, he felt exposed, being by myself and unarmed and faced with the unknown. After a couple of minutes he lower lower back to Gimlin, the pair captured their horses, which had run away in terror, and made plaster casts of the footprints the creature had left in the back of. Almost two weeks later, community taxidermist Robert Titmus visited the internet website on line and made plaster casts of a chain of large footprints. Since it had rained among the film being shot and Titmus arriving, those prints have been likely no longer made in some unspecified time in the future of the filming but afterwards, suggesting that the beast got here decrease lower back.

While Patterson and Gimlin portrayed the come upon as an accidental one, skeptics factor out that Patterson emerge as an avid Bigfoot hunter in advance than he ever took the pictures and were on numerous searches for the creature, finding its tracks or maybe writing a ebook about it. The had been, in reality, on a Bigfoot hunt that day. While that doesn't lessen rate the possibility that they stumbled upon what they were searching out, it gives a few doubt in the minds of those already inclined to doubt.

Also, primatologists aspect out that the creature does now not walk like each regarded primate, but alternatively a person in a monkey match. It's tough to mention, however, given that Bigfoot is, of course, no longer any stated primate however as an possibility some element from the unknown, so it isn't always viable to make certain how it'd stroll.

The affair have become similarly muddled in 1999, while Bob Heironimus issued a press release announcing that he had worn an ape in form as a manner to make the movie. More damning proof got here from Philip Morris, owner of Morris Costumes, who in 2002 claimed that he had supplied Patterson an ape in shape to use within the movie, and endorsed him on a manner to modify it to make it appearance greater huge, which consist of having the actor placed on soccer pads underneath the healthy and hold sticks in his palms to make the hands appear longer. This will be damning testimony except for the fact that Bob Heironimus, at the same time as asked in an interview if the in form were padded, said it emerge as not. Since Heironimus does not have the bodily proportions of the creature in the movie, he would in all likelihood have wanted the improvements Morris mentions.

An unusual function emerged inside the Patterson-Gimlin movie whilst cryptozoologist M.K. Davis analyzed it the use of pc enhancement. He discovered that in the woods in advance of the Bigfoot there was a darkish decide that changed into inside the form of every other Bigfoot. It's in reality now not viable to look with the bare eye on the movie itself, but the usage of freeze frame and turning the color into black and white with a view to decorate the assessment, it in all fairness smooth. If Patterson and Gimlin wanted to fake a Bigfoot film, they wouldn't conceal a 2d man in a monkey healthful deep within the shadows of the timber in which no character could see him. Neither Morris nor Heironimus point out a 2d person dressing up in a wholesome and posing for the digital camera.

The controversy continues, as do the sightings. I for my part comprehend a person who claimed to have an come across with a Sasquatch in a miles off stretch of woods in western Missouri about 15 years in the past. He have become tenting on my own, generally an invite for journey, and after a pleasant day of hiking, he rolled himself up in his slumbering bag and permit the severa sounds of the forest lull him to sleep.

Late that night time time a noisy thud awoke him. The thud shook the floor, and his instant effect as his eyes snapped open became that he had heard the footfall of some huge beast.

His subsequent affect end up extra ominous—the forest become definitely silent. No owls hooted, no cicadas chirped, no flies buzzed. There turned into no sound in any respect. It changed into as even though all of the creatures of these lush woodlands were protecting their breath, hoping that something had made that high-quality footfall would possibly skip them by way of. My buddy held his breath too.

Then he heard a valid that grew to grow to be his blood to ice. There emerge as a loud crashing via the undergrowth, a snapping and breaking of branches and timber as a few component massive moved away, and a sequence of loud, thudding footprints that shook the floor and made my pal curl up in terror.

Then the creature changed into lengthy past. It took a long term for the regular sounds of the forest to come again again lower returned, and longer still in advance than my pal have been given any sleep.

What occurred? It isn't feasible to say. It have become overdue at night time time, darkish, and he turn out to be half asleep. Perhaps he imagined all of it. Perhaps this become clearly a few one of a kind fireplace camping story to whilst away the night time. Perhaps he modified into wrong.

The truth that this guy come to be a lifelong stoner did no longer help his credibility. He moreover claimed to have seen a sparkling pixie dancing inside the woods, however I won't stretch the reader's credulity to the breaking detail by way of way of pertaining to that story.

Bigfoot appears to have network variations or subspecies. In Missouri, the Momo ("Missouri Monster") is a seven foot tall apelike being with black hair and an nearly insufferable stench that lingers lengthy after the creature had departed. This isn't some factor that's normally noted in Bigfoot sightings, and my camper friend said nothing about a awful smell. Unlike most Bigfoot tracks, Momo's big footprints display it to have first-class three toes. It additionally seems now not to have a neck, while the famous Patterson-Gimlin movie simply shows Bigfoot turning his head to observe the returned of at the digital digital camera.

In Oklahoma in 1970 and 1971, farmers had to attend to the Abominable Chicken Man. No one ever observed it, however the tracks regarded find it irresistible become taking walks on all fours, using its hands as forefeet like an ape. Lawrence Curtis, director of the Oklahoma City Zoo, went on report bringing up that the prints looked just like the ones of a primate, however said no primates lacking from his care. This creature changed into mainly violent, ripping aside bird coops and devouring the chickens inside.

In Asia, there can be of route the Yeti or Abominable Snowman of the Himalayas. This creature, who appears to be much like Bigfoot besides for being white, lives within the big, desolate heights of the area's tallest mountain chain. Just what the Yetis devour up there can be anyone's bet. Numerous mountaineers have noticed the Yeti or its massive, unique footprints.

The legend of the Yeti dates to many centuries in the past, from the pagan era in advance than the vicinity embraced Buddhism. It emerge as believed that a wild man lived in the mountains, his blood being an important element in numerous magical rituals. The Lepcha humans worshipped this "Glacier Being" due to the fact the God of the Hunt. When Buddhism modified into brought, the new religion delivered in lots of close by ideals, along with that of the Yeti. Several Tibetan

Buddhist monasteries maintain Yeti scalps or even preserved Yeti fingers.

The Yeti first came to the eye of Western era while in 1832 the Journal of the Asiatic Society of Bengal posted an account via English naturalist Brian Houghton Hodgson about his excursion to the Himalayas. While up inside the mountains analyzing the various fauna, maximum of which had however to be scientifically defined, his publications knowledgeable him that that they had observed a big furry creature that fled once they came into view. Hodgson disregarded the idea that this became a few new species unknown to technological knowledge and said that in his opinion it grow to be an orangutan that had arise from the lowlands.

Perhaps the primary Western eyewitness of the creature itself changed into the photographer N. A. Tombazi, who wrote of his reviews in the 1925 ebook Account of a

Photographic Expedition to the Southern Glaciers of Kangchenjunga within the Sikkim Himalaya. One day, at the same time as approximately ten miles from the Zemu Gap at an elevation of about 15,000 feet (4,572 m), his porters known as him out of his tent to witness a high-quality sight. About 250 yards (229 meters) away and further down the slope stood a weird determine silhouetted toward the evident snow. He wrote, "Unquestionably, the determine in define modified into precisely like a person or women, on foot upright and preventing now and again to drag at some dwarf rhododendron wooden. It confirmed up dark in the path of the snow, and as a long way as I need to make out, wore no clothes." The determine have become seen for about a minute in advance than it moved off out of sight. When Tombazi and his organization made it to the aspect wherein they located the creature, they discovered its prints. "I examined the footprints that have been simply seen at the floor of the snow. They had been comparable in form to those of someone, but best six to seven inches lengthy by way of the use of four inches extensive on the broadest part of the foot. The marks of five excellent toes and of the instep had been perfectly clear, but the trace of the heel modified into vague, and the little that is probably seen of it appeared to narrow down to a degree. I counted fifteen such footprints at everyday durations beginning from one-and-a-half of of to 2 ft. The prints have been genuinely of a biped, the order of the spoor having no trends some issue of any possible quadruped. Dense rhododendron scrub avoided any in addition investigations as to the direction of the footprints, and threatening weather pressured me to renew the march. From inquiries I made a few days later at Yoksun, on my pass again adventure, I gathered that no man had lengthy long long past inside the direction of Jongri considering the truth that the start of the only year."

The Yeti and its tracks were seen numerous times ever due to the fact, consisting of via famous human beings such as Sir Edmund Hillary and Tenzing Norgay, the number one men to scale Mt. Everest. In 1953 they stated seeing many big tracks they could not pick out. Norgay said that his very personal father had seen the Yeti, regardless of the truth that he himself doubted its existence.

Back in 1994, this creator went hiking inside the Annapurna Range of the Himalayas in Nepal, headed for the Annapurna Base Camp and modified into excessive inside the mountains at an altitude of approximately 10,000 toes (3000 meters). While crossing a big, bowl fashioned valley with hard, rocky peaks to both side, at about midmorning some huge footprints inside the snow had been observed. They measured about 20 inches prolonged and approximately 10 inches good sized, with huge impressions within the the the front that seemed like ft. The tracks led some distance from the center of the valley to considered one in every of its edges.

Tromping thru the deep snow and following the tracks delivered about a tangle of rocks close to the bottom of the cliffs at the valley's side. The stride of the animal that left the ones tracks have become strangely quick for the scale of the foot, and the purpose for this have end up smooth upon achieving the coloration of the cliffs, a part of the valley that had not but been touched with the useful resource of using the sun. Rather suddenly, the course of prints shrank in length to that of some small animal; the footprints went from twenty inches to four inches, and the toe prints end up narrow impressions of claws. In impact, the solar, which is strong within the ones excessive mountains, had partly melted the snow, broadening and enlarging the prints to expose them from a few sort of animal like a fox into some thing paranormal.

While this can give an explanation for among the Yeti tracks humans have seen in the snow, educated observers have decided tracks they couldn't so effects supply an reason of, and personal observations additionally cannot provide an reason for the eyewitness sightings or the centuries-vintage way of life that there may be surely a few shape of creature inside the super mountain fastnesses. On the same trek, there was a Sherpa who claimed to have visible the Yeti within the equal place of the in the end disappointing tracks. He said he changed into springing up a trail while pretty he noticed the creature sitting on a rock not some distance in advance. He stated it "appeared like a person," modified into now not specially big, however covered in white hair. As soon because the creature noticed the Sherpa, it leapt up and ran off with a loping stride.

Unlike with many cryptids, there are several scalps, toes, and arms that supposedly belonged to the Yeti. Some of those had been subjected to medical assessment, collectively with DNA checks. The first come to be finished on the so-called Snowman Expedition of 1954, sponsored thru the British newspaper the Daily Mail. This tour have become almost tailor made not to discover a Yeti. Dozens of photographers, filmmakers, mountaineers, naturalists, and reporters had been supported via numerous extra dozens of porters and courses. This small military made a big quantity of noise clambering alongside the mountain passes seeking out a skittish creature that tends to run away at the sight of a single human.

The tour did discover a few tracks the scientists inside the group couldn't discover, and they spent lots of their time journeying Buddhist monasteries to check purported Yeti factors. At the Pangboche monastery within the Imja Khole valley of Nepal they tested a Yeti scalp the priests had preserved. The take a look at of a number of the hairs from the scalp became led thru Professor Frederic Wood Jones, a Fellow of the Royal Society and a global-renowned expert in comparative anatomy. His findings had been inconclusive. He couldn't emerge as privy to the species of the animal, notwithstanding the fact that he recommended they were from some hoofed quadruped and not a simian. He moreover belief the piece of hairy flesh changed into from the shoulder of this animal, now not the scalp, due to the truth no animal has a ridge walking from the forehead across the pate and to the once more of the neck. No regarded animal, at any price.

In 1959, some supposed Yeti feces have been analyzed and discovered to incorporate an unidentifiable parasite. Since parasites are enormously species particular—the not unusual head louse (Pediculus humanus capitis) is simplest observed in humans—some cryptozoologists purpose that the host species is likewise unknown to era.

An in addition flimsy little bit of proof is the Pangboche hand, which got here from the identical monastery because the scalp Dr. Jones analyzed. One of the hands emerge as stolen from the hand with the aid of cryptozoologist Peter Byrne in 1959 and none other than well-known actor James Stewart smuggled it in another country in his luggage. Several DNA tests have been achieved, with results various from "near human" to "human." This does no longer, but, brush aside the concept of the Yeti, when you keep in mind that it can be a close to relative of our species. Some meant Yeti hair have become analyzed by using Oxford Brookes University in 2008 and decided to belong to a Himalayan goral, a form of mountain goat.

Of path, faux Yeti scalps and hands do now not prove that the Yeti itself is a fake, and cryptozoologists hold to theorize approximately this elusive humanoid. Some accept as true with the severa Bigfoot and Yeti creatures might be a surviving populace of the Gigantopithecus, a genus of ape that lived from approximately one million to 3 hundred,000 years ago. Others country it's miles a Paranthropus robustus, an ancestor of modern guy that lived from 2 to at least one.2 million years inside the beyond, or probably some previously unknown hominid.

Chapter 6: ABCs and other OOPAs

Sometimes human beings don't see monsters, however ordinary animals that shouldn't be there. Cryptozoologists generally name these Out of Place Animals (OOPAs), the maximum common kind being Alien Big Cats (ABCs).

In the us there were severa reviews of untamed kangaroos. One early but pretty demanding record comes from 1934, even as Reverend W. J. Hancock of South Pittsburgh, Tennessee, noticed what appeared like a large kangaroo attack and kill a canine, that is uncommon conduct for a kangaroo. The loveable creatures are purported to be herbivores. It's additionally unusual for kangaroos to loaf around in Tennessee. Other witnesses also noticed the animal over the subsequent numerous days, earlier than it vanished as mysteriously because it appeared.

While the ones damn Yankees up north might probable brush aside this tale due to the fact the wild imaginings of bug-eyed rednecks, kangaroos appear within the Northern states too. In 1974 in Chicago, police had been known as to intervene whilst a person stated a kangaroo popularity on their the the front porch. The police observed the creature but had been no longer capable of seize it. This prompt a wave of information in Illinois, Indiana, and Wisconsin that lasted approximately a month before the animal or animals disappeared.

More commonplace are big pussycats recounted to cryptozoologists as Alien Big Cats. Alien Big Cats are determined in places wherein no big felines are presupposed to exist, at least not in the contemporary era. The United States has had severa sightings of this type of Out Of Place Animal.

One of the prime breeding grounds for Alien Big Cats, and in reality all styles of cryptids, is the Appalachian Mountains, a 1,500 mile (2,four hundred km) mountain variety stretching from Canada to the southern United States. Not first rate is it sincerely one in all the maximum important mountain levels in North America, but it's also the oldest, having been usual some 480 million years inside the beyond. Time has worn down the peaks, leaving the tallest first-rate 6,684 ft (2,037 m) above sea degree, however this has additionally made for a wealthy topsoil and dense wooded region cowl that might disguise any sort of unfamiliar beings.

The maximum usually visible is the Appalachian Black Panther. Panthers are not presupposed to exist within the United States. There are black panthers in Asia and Africa (in truth they're melanistic leopards) and black jaguars in Central and

South America, but no such creature is meant to live in North America. Cougars do exist in North America, however have supposedly been extinct in the Appalachians for added than a century. Could the Appalachian Black Panther in reality be an extremely good cougar that has melanistic fur?

The large type of sightings indicates that this is the case. Numerous farmers and hikers have referred to huge, black cats, too big to be an outsized housecat. There have moreover been numerous photographs and films of the beast. To in addition beautify this speculation, we need simplest aspect out numerous sightings over the years of commonly coloured cougars in the Appalachians. While no specimen has been stuck, there is strong proof that this Alien Big Cat does honestly exist. This is thrilling territory, in which cryptozoology comes tantalizingly close to mainstream technology.

England is specifically wealthy in ABC sightings, possibly due to the English love of cats. One of the maximum well-known is the Beast of Bodmin Moor. In a moor in Cornwall, many human beings have sighted what looks like a black panther. The Ministry of Agriculture, Fisheries and Food has recorded more than 60 sightings of the beast, the earliest dating from 1983. Numerous photographs have been taken of this creature, but they may be generally fuzzy and from an prolonged distance and shortage any landmarks that might help determine the size of the creature being photographed. Because of this, scientists usually brush aside the sightings as misidentified housecats. Why a cat-loving Englishman would be now not capable to differentiate among a cat and a panther, why those sightings are so commonplace and have been occurring for decades, and why such numerous cattle are being ripped aside inside the area are the diverse many questions which might be left unanswered.

The elegant cause for OOPAs is that a zoo, circus, or private collector allowed the animal to interrupt out. The hassle with this is that everybody who's legally being worried for such an animal might make an declaration about the break out, no longer top notch to guard their pals, however in the hopes of getting their high priced animal decrease again. Cryptozoologists have records masses of OOPAs wherein no such assertion end up made. So barring a giant illegal community of wonderful animal creditors who're strangely careless, this remains a crucial thriller.

Chapter 7: Too Strange For Classification

Some evaluations of cryptids defy categorization. These are typically simply considered one of a kind creatures, and we must be thankful that they're certainly considered one in all a kind. They moreover will be predisposed to be decided in a completely localized location, which all once more is a high-quality element.

The Goatman of Maryland is that this form of unclassifiable cryptids. Said to hang-out Prince George's County in Maryland, this is one of the many close by beasts which is probably determined everywhere within the global. The Goatman appears to be a person included in goat hair, or as a few witnesses record, it has the pinnacle frame of a regular guy and the lower frame of a goat, bringing to thoughts the satyrs of Greek mythology. It prowls the woods and backyards at night time time looking puppies, which is probably its favored food.

The Goatman is meant to be a scientist who labored in the close by Agricultural Research Center. He modified into walking on some forbidden experiments even as, in the incredible B film manner of existence, a few factor went incorrect and he transformed proper into a half of-man, half of-beast. This grew to become him right proper into a madman, or a mad Goatman, and he fled to the deep woods, dwelling in a shack and most effective emerging to consume dogs and scare necking teenagers on Lover's Lane. Sometimes he beats on their motors with an awl, an extraordinary example of a cryptid the use of a tool. Sightings commenced out in 1957 and he is meant to have murdered severa hikers in 1962. He continues his midnight prowls to in the meanwhile, his axe on the equipped.

This sounds similar to the fearsome Bunny Man sighted in 1970 and 1971 in Washington DC. As the name implies, this regarded like a human sized, bipedal bunny. You chuckle? You wouldn't giggle if he grow to be the usage of an awl to spoil your vehicle or reduce down the permits of your home, because the Bunny Man modified into stated to have executed. Of route the Goatman and Bunny Man ought to in reality be psychotics sporting stupid costumes and strolling around with axes, however that's no longer any more reassuring than the opportunity of a few monster at the free.

There are Birdmen too, and those have an extended statistics than each the Goatman or Bunny Man. On September 12, 1880, a letter discovered out within the New York Times stated that "many dependable men and women" had visible a humanlike creature flying over Brooklyn "and they all agree that it emerge as someone engaged in flying inside the course of New Jersey." He modified into described as a "guy with bat's wings and superior frog's legs," some issue meaning. Not fantastic modified into he flapping his wings, but shifting his legs like a frog swimming thru water.

One might also want to brush aside this as the letter of a crank, posted via an in any other case extreme newspaper for the fun of it, but reports endured. In 1948, a Birdman modified into noticed over Oregon flapping its big silver wings. In 1953 he took a flight over Houston, alternatively in 1961 over West Virginia, and once more in 1966 over Greenville, Mississippi.

Could this actually be severa generations of mad scientists showing off their invention? There have been many experiments the use of flapping wings, specifically earlier than the Wright

Brothers developed the primary powered aircraft, however no one has showed robust sufficient to get themselves off the floor, not to mention fly immoderate over towns. Humans in reality aren't made for winged flight.

The United States isn't the only america of the united states that's domestic to a number of those localized creatures. In Cornwall, England, there's the Owlman, additionally known as the Owlman of Mawnan, due to its many appearances inside the village of Mawnan inside the past due 1970s.

The creature first seemed in 1976, even as it have come to be sighted in 3 separate incidents, all with the useful resource of extra younger or teenage ladies at night time. It regarded at night time, became approximately six ft tall with big wings of silvery gray feathers, legs that bent backwards like a fowl's, and pincers for toes, which one witness defined as just like a crab's claws. Other top notch capabilities were a big, fanged mouth and glowing pink eyes. The creature emitted a noisy hissing sound that nervous the ladies.

There were no greater sightings that year, and people disregarded it because the overactive creativeness of schoolgirls staying out too past due for their private appropriate. But then in 1978 three French alternate university university college students also noticed the creature, and that identical one year the Owlman turn out to be noticed through some other close by girl. The next sighting turn out to be with the resource of the use of a greater youthful couple in 1989. Two extra sightings got here in 1995, one through an American student. It's exciting whilst a neighborhood monster is suggested via the usage of someone new to the vicinity, as they're plenty a great deal less probably to had been the difficulty of unconscious concept thru the place's folklore. All the opinions agree in the identical latest data, particularly the sparkling red eyes and the horrifying hiss the creature emitted. Whatever it emerge as, the fearsome Owlman and its sparkling crimson eyes have in no way been sighted once more.

A bit more viable than the Owlman and his American cousins is some one-of-a-kind close by creature, the Beast of Dean. This is positioned within the Forest of Dean in Gloucestershire, England. It come to be first observed numerous times in 1802 and emerge as described as an unnaturally huge wild boar that permit out an "unearthly roar." It had excellent tusks which it used to knock down wood. It modified into visible for plenty many years but the sightings subsequently petered out. The Beast of Dean have become forgotten to all except a few folklorists until 1998, at the same time as it reappeared. The two witnesses said it appeared just like a boar besides that it have come to be the size of a cow. It made each other look on November 7, 2005 and modified into advocated inside the close by paper, The Citizen.

The Beast of Dean seems like definitely an outsized boar, besides for the atypical fact that there have to were as a minimum of them, one within the early nineteenth century and some one-of-a-kind almost 100 years later. Could a few component inside the Forest of Dean be making an occasional boar broaden to a big period?

We have already seen that the Appalachian Mountains are domestic to many ABCs, however the Appalachians also are domestic to 2 fearsome variations of the wolf which can be too outstanding to be considered Out Of Place Animals.

The Dwayyo looks as if an outsized wolf that every so often stands on its hind legs to a top of as heaps as nine toes (three meters). It we ought to out a terrifying howl at night time and has been visible by way of manner of a number of humans, along aspect park rangers. It should be handled with caution, for it's miles stated to be the perpetrator of the severa canine and livestock mutilations in the vicinity. Another canine critter is the Snarly Yow, that may be a huge black canine with a red mouth and huge fangs that has been visible at the same time as you endure in mind that Colonial times. This creature can be more paranormal phenomenon than cryptid, as it's miles stated to change colour and length and people who've shot at it said their bullets handed right through. There are numerous opinions of "Demon Dogs" from Europe, mainly England, and the Snarly Yow also can truly be a close-by continuation of that human beings legend.

Another cryptid that appears to stick to a totally localized vicinity is the Jersey Devil, which lurks within the Pine Barrens of southern New Jersey. Why it want to first-class be on this kind of small place is a mystery, thinking about the truth that it's far stated to have wings and will as a end result fly everywhere it favored to.

The Jersey Devil have become first visible in 1909, on the identical time as from among January 16 to 23 there was a wave of sightings of what witnesses stated it regarded like a kangaroo with a goat's head; leathery, batlike wings; antlers like a stag; and cloven hooves. The monster had small palms with little clawed palms and a forked tail. It may also need to fly, changed into frequently competitive, attacking cattle or maybe humans, and emitted a horrible blood-curdling scream.

Its eyes are stated to glow a colourful purple. A blood-curdling scream and purple eyes are taken into consideration crucial for any self-respecting monster.

The reviews have been so huge that faculties closed or even many adults stayed domestic, specially the mill human beings within the Pine Barrens wherein the creature became stated to reside. The Philadelphia Zoo supplied a $10,000 reward for the critter, however no one should produce one no matter several teams of armed men going out into the woods in search for it. One hopes no character have been given shot by means of coincidence. I changed into nearly shot by means of manner of a few drunk hunters whilst hiking in Arizona and I look no longer some thing just like the Jersey Devil. These hunts did produce several cloven tracks that would have been the beast. The prints no longer most effective appeared everyday, they did ordinary topics. They often went proper via fences with out breaking them, as though the constraints weren't there, and snow-blanketed rooftops all around the vicinity display signs and symptoms of the Jersey Devil's passing. While most of these tracks have been decided in rural regions, some seemed within the middle of Philadelphia as properly. Devil hunters determined the tracks but by no means came upon the monster. A crammed Jersey Devil did display up on the circus sideshow circuit that twelve months, but it became out simplest to be a kangaroo with some fake wings glued to it.

The small town of Camden have emerge as a middle of sightings. Police shot at it to no impact, and it attacked a trolley vehicle and a social membership. The Jersey Devil regarded to have more bark than chew, however, due to the truth even as a close-by woman saw it attacking her canine, she swiped at it alongside side her broom and it flapped away.

Reports of the Jersey Devil died down after about in step with week, however by no means definitely went away. It has been seen often for the beyond one hundred years. One tale recounted thru the Asbury Park Press expenses an unnamed guy who noticed the creature in 1981. He stated the beast stood approximately six toes tall, had top notch teeth in a horselike mouth, and fur sooner or later of its frame. Strangely, he stated it had 3 toes on each foot, no longer cloven hooves like maximum memories. If someone turn out to be going to make up a sighting, it stands to reason that he need to stay with way of life.

Jersey Devil drawing from 1909

Some say that the beast has been round loads longer than thinking about that 1909. The real Native American citizens of the area, the Lenape, referred to as the area "Popuessing," this means that that "Place of the Dragon." There's additionally a legend that the Jersey Devil is the son of an 18th century witch. The tale is going that Deborah Leeds gave shipping to 12 everyday youngsters, and while in the yr 1735 she have been given pregnant for the 13th time, she admitted that the satan became the daddy. She gave start one stormy night time time to what regarded to be a wholesome, everyday boy. Then, before definitely each person's eyes, the infant converted into the Jersey Devil. Its first sufferer modified into the midwife! Once the monster had a few meals in its belly, it flew up the chimney and has been terrorizing the location ever due to the fact that. Locals often name the Jersey Devil the Leeds Devil, and a number of the

1909 newspaper opinions did in order nicely.

Folklorists keep in mind that this story became only made up in the 20th century and cryptozoologists, who are susceptible to push aside paranormal elements for the cryptids they check, have a tendency to agree. A query cryptozoologists must ask with those localized creatures collectively with the Jersey Devil or Goatman is why simplest one is ever seen. If there is no breeding populace, they could not live on for good-bye. Could those creatures be a few form of wild mutation in area of an actual species?

Chapter 8: Early Explorers Before Columbus

Columbus Day stays determined for the duration of America to commemorate Christopher Columbus' discovery and exploration of the New World. However, it's been extended everyday that terrific Europeans determined America first. Archaeologists take shipping of as actual with that the Bering Strait became crossed by means of the usage of way of Asians spherical thirteen,000-16,000. The sea diploma became decrease returned then and there was a land link between Alaska and Russia. It is thought that Native Americans' ancestors crossed the bridge and spread slowly via the New World.

This idea has a problem due to the reality there are developing numbers of archaeological web web sites that date decrease returned to masses in advance times. Monte Verde is positioned in Chile and dates again to fourteen.800 years within the beyond. Meadowcroft Rockshelter, Pennsylvania, dates lower lower back to 19,000 years. Pedra Furada, Brazil, dates decrease once more to over 60,000 years. These net web sites are not well preserved and cannot be radiocarbon dated. The stone gadget decided there also can be herbal chipped rocks.

While archaeologists are though seeking to discover the lacking hyperlink, there may be developing proof that Asians may additionally want to have reached the New World through the use of themselves with out the Bering Strait Landbridge.

Archaeologists are beginning to agree that the Americas have been colonized earlier. What approximately after that? Reminding my Eurocentric trainer, which European changed into the primary to reap America? There is concrete evidence to show that Leif Eriksson, at the side of different Norsemen, made it to New World within the yr one thousand AD. L'Anse aux Meadows, a Newfoundland archaeological internet site on-line, has been preserved and observed stays of Viking homes, workshops, artifacts, and specific devices.

Are there special cultures that might have reached North America? The winning winds and ocean currents of the Atlantic recommend that a deliver can be without trouble carried from Western Europe to the Caribbean. There, the currents alternate to keep the vessel up the east coast to North America.

From there, it'll tour over again to Europe via the Atlantic. These currents were additionally used by ships inside the age in advance than sail. Even a primitive vessel have to have reached the vacation spot if it had enough sources and revel in, further to exact fortune.

But ought to actually everybody have completed that? There must were many historic ships that had been blown off course and ended up in New World.

It isn't clean if the organization had been however alive at that factor. The essential problem is whether or not the voyage changed into made on reason and if the institution made common voyages most of the Old World, and the New.

Some researchers in New England don't forget that they have proof that ancient cultures may have completed this.

They element out unusual stone systems and incriptions which seem like in historic Old World language. These incriptions, they claim, were now not made with the resource of the Native Americans or the early colonists.

There are many small stone chambers scattered ultimately of New England and the northern Mid-Atlantic States that puzzle researchers. They are made the usage of the dry stone method.

This consists of honestly placing rocks on top of each unique with out a mortar. It is also the same method that have become used to make the ever present stone boundary partitions inside the region. Most are rectangular or rectangular in shape (although a few have a round interior), but others are very spacious.

The longest I without a doubt have seen measured 30 feet in length, 10 toes large and eight feet excessive. Many have roof slabs or lintels that could weigh up to numerous loads. Although maximum are set on a hill or with earth spherical their aspects, a few are. These chambers are scattered within the path of the place. Many more were destroyed thru "development" inside the 20th century, whilst the land changed into stripped of plenty of its historic relics.

They aren't in all likelihood to had been made with the aid of Native Americans who in this location did not construct in stone. These chambers can be decided inside the rocky uplands wherein Native Americans in no way lived completely.

They also can have traveled to the highlands to are attempting to find or accumulate but they favored to stay inside the well-watered lowlands wherein they'll will be predisposed to their plants.

This is wherein the pre-Columbian New England Native Americans made their eternal homes. These uncommon-looking, stone structures have been claimed through historians to be Colonial root cellars that have been used for meals garage.

However, no longer all researchers are in agreement. These researchers take delivery of as real with that they may be remnants of an ancient civilization which colonized the New World centuries in the past.

Columbus. This principle is primarily based totally on some of presumptive inscriptions in historical languages. Most extensively Ogham, an bizarre script that modified into decided inside the British Isles, Ireland and across the 4th to 9th centuries AD.

Ogham may be defined as a series of lines which might be either at once or slanted, quick or extended, alongside or one thing of a huge line. One line on the handiest component of the important divide is a "B" and five slanted strains through it are an "R." Ogham have become made up of heterosexual or slanted strains. A 14th-century Irish manuscript referred to as the

Book of Ballymote contains the key to nearly a hundred Ogham variation alphabets. However, a few Ogham version alphabets have been determined in early Irish manuscripts.

Barry Fell, an beginner epigrapher and marine biologist, modified into the first to signify that some of the gashes on New England's stones had been Ogham inscriptions. He decided hundreds of them in New England's root cellars, herbal stone outcroppings, in addition to stones in stone partitions made from dry stone that he claimed were quarried from antique rooms after which reused.

Fell and different epigraphers translated masses of those inscribeds. They determined that most are non secular in nature and phone antique Celtic gods like Bel (written in Ogham, with one decrease observed via extra). They do not forget the foundation cellars are temples. Ogham script can be used to install writing many languages, collectively with early Celtic and Iberian Celtic that have been utilized in historical Spain, Portugal, and different worldwide locations.

Others have moreover decided astronomical alignments inner a few chambers. The Vermont Calendar I and Calendar II web websites align with the Winter Solstice Sunrise.

The door of the Calendar I stone chamber is aligned to a close-by ridge. The sun rises on a saddle at the ridge at dawn on the Winter Solstice.

The solar rises on the Calendar II net website on line thru a notch created by way of the use of hills close by, this is truly seen from the chamber's returned. Other stone chambers also have alignments with one-of-a-type vital instances in astronomical calendar. Some Ogham inscriptions seem like associated with those astronomical activities.

Inscriptions additionally may be positioned in extraordinary languages. A stone slab with ordinary glyphs is positioned within the Bourne Historical Society, Massachusetts. They are believed to be Native American petroglyphs.

However, Fell claimed that the script became an early Celtiberian script. The Iberian Celts wrote in their very very own language, not Ogham. The inscription changed into translated through Hanno as "A proclamation for annexation." This is how Hanno takes possession. Hanno changed proper right into a Phoenician explorer, who travelled alongside the west coast Africa's in the first 1/2 of the sixth century.

BC.

A slab of bedrock close to Dighton in Massachusetts is every different interest. It has oddly overlapping carvings. One of the carvings seems like a human parent and some are geometric shapes. He stated, "It is noted that an vintage Indian legend has it that there has been a timber residence with guys from each distinct u . S . That came up the river Assonet and defeated the Indians. They then killed their Sachem. Many trust the figures are hieroglyphic. The first represents a deliver with out mast and a easy damage sturdy at the shoals. The second represents a head of land. It may be a cape or a peninsula.

Carl Christian Rafn, a Danish scholar, idea it became a Norse inscription. He translated it into "Thorfinn, his 151 friends, took possession of this land" in 1837.

Edmund Delabarre, a 1912 author, wrote that the Latin abbreviation of the inscription test: "I, Miguel Cortereal 1511." Cortereal, a Portuguese explorer, disappeared from the East Coast in 1502.

Dighton Rock

Some translations claim that the inscriptions were Ogham, Phoenician or Chinese. You can now see the rock in Dighton Rock State Park and puzzle over the carvings.

The American Ogham precept has many troubles. First, there's no evidence of Ogham in Europe previous to the 4th Century AD. Second, it modified into by no means used drastically.

There are just a few hundred Ogham inscriptions left, maximum of which can be placed in Ireland or Wales. None were placed in Iberia or a few specific vicinity colonized by using the Phoenicians. Dr. Fell desires us to bear in mind Ogham is from severa centuries earlier than that, and for why.

It turn out to be used by Celtiberians to jot down down the Phoenician language whilst both have been written in it.

The Phoenicians and Celtiberians had already written their very own languages. The script is likewise difficult. Ogham consists of a sequence right away or slanted traces. This manner that any herbal cut within the rock, whether or not or now not or no longer it's far from roots, ploughs or manmade scrapes, should likely look like an Ogham script.

The critical line in the New England examples is normally missing. This makes it difficult to choose out individual letters. The primary line in Old World examples is usually both inscribed or crafted from the threshold the monument.

Yet, there can be however uncertainty even amongst archaeologists. It begs the question, if such a variety of alleged historic inscriptions are placed in New England, is it possible that a number of them can be proper? What is the motive for apparent astronomical alignments?

Alternative theories are not restrained to stone chambers. There are many examples of huge boulders atop smaller rocks scattered across the us. These are believed to be dolmens.

This is a shape of Neolithic megalithic burial chamber that can be decided in Europe, Asia and Africa. The earth that used to cover those tombs has typically eroded away leaving uncovered stones, as visible in the New England examples. This concept has a problem.

True dolmens are built with large help stones to permit for grave items and bodies. New England examples have smaller stones. Some of these that I've visible wouldn't had been huge sufficient to hold even one frame.

These dolmens are believed to were created via the taking flight glaciers on the stop of the final Ice Age, dumping rocks as they moved. This is why the vicinity is stuffed boulders and small stones which have been left inside the lower again of.

Sometimes, inside the route of the deposition way, big rocks would decide smaller ones and create a glance that resembles a dolmen.

Others sites may be difficult to miss. The maximum famous is Mystery Hill in Salem, New Hampshire. It's additionally referred to as America's Stonehenge.

Locals were questioning about the weird series of stone shelters Locals were wondering approximately the weird collection of stone shelters acre internet internet site on line for greater than 100 years. One area consists of a shape that looks to be an altar.

It has a channel that runs round its outside, which some be given as real with is meant to empty the blood of the sacrificial victim. Under the altar is a tunnel and a small hollow known as the "Speaking Tube", which might be hidden from view thru people outdoor.

The tunnel's noises sound almost as although the sounds are coming from the altar. This is the "Oracle Chamber", as America's Stonehenge proprietors go to it.

America's Stonehenge"

Researchers have located severa astronomical alignments in the stones. The "Watch House," a chamber positioned on the net site on-line's aspect, has a again wall manufactured from stone that includes quite a few quartz. It is flanked thru the use of a window that lets in the growing sun on February 1. This makes the returned wall seem to shimmer. The Celtic calendar considered February 1 an critical date, as it became the day of Imbolc, which celebrated the begin of spring. Some of the larger popularity stones are aligned with the solstices or equinoxes in different regions of the net page.

William Goodwin sold the website in 1937. He renamed it Mystery Hill whilst he opened it as tons as tourists that year. Goodwin believed that it have become the home of Irish clergymen fleeing the Vikings.

Some structures endure a superficial resemblance with the dry stone systems utilized in medieval priests of western Ireland. Goodwin rebuilt the website on line with the resource of shifting stones and rebuilding the collapsed systems within the equal way that he believed they have been constructed.

Later studies disagreed collectively together with his interpretation. Traditional archaeologists be given as real with the systems date returned to the 18th or nineteenth century and are consistent with storage shelters constructed with the resource of New England farmers.

The sacrificial rock is genuinely a lye-leachingstone, which became used to extract lye from timber and ash to make cleaning soap. Barry Fell claimed that he discovered Ogham inscriptions on the net internet site, further to Phoenician writing and Iberian writing. Modern excavations have located out Native American artifacts and capabilities like fireside pits that date lower lower returned to 2,000 BC. However, those appear to be from an in advance use of Mystery Hill and it isn't probably that Mystery Hill have come to be built with the aid of way of Native Americans.

The Newport Tower in Rhode Island is every different unusual structure that has drawn opportunity archaeologists. It is usually believed to be an 17thcentury windmill. However, a 1677 document refers to it only as a windmill and there's no cause for it to had been some thing else. Two excavations were made on the internet site on line and radiocarbon dating of the mortar used for cementing the stones gave a date amongst 1635-1698.

The Newport Tower, photos via Matthew Trump

Although it's miles obvious that the Newport tower emerge as constructed in 1898, distinctive researchers receive as authentic with it to be more than a mill.

William Penhallow, an astronomy professor, tested the tower's home home home windows that have been no longer nicely located and decided they had been aligned with many astronomical activities. The west window is positioned so that the putting sun shines thru it on the summer time solstice. It moreover falls on an inner wall niche. The east and west windows are aligned on the southernmost aspect during the minor lunar standstill. This is just like the sun's solstice in that the immediately at which the moon sets, which adjustments each day, appears to stay consistent on the horizon for severa days. These sun and lunar standstills have been utilized by ancient people to music the time and assist them plan when they need to plant and harvest vegetation. The test of stone circles and other megalithic structures through archaeoastronomers.

Europe has decided alignments that correspond to lunar standstills. Gavin Menzies, in his e book 1421: The Year China Discovered America claims that the observatory come to be built through using using the Chinese admiral Zeng.

Others, more esoteric, advocate that the tower has mystery numerological importance or is a map displaying historic civilisations made in stone. The Newport Tower attracts curious traffic, no matter whether or not or no longer it's miles a smooth milling device or the crucial thing to esoteric information.

Many Old World artifacts from the Old World, including Celtic swords and Roman cash had been positioned in New England. However, they may be nearly constantly not decided at some point of professional archaeological investigations. The Maine Penny, a small silver coin from Norway that dates once more to 1065-1080, is the handiest pre-Columbian artifact that has been confirmed in New England.

A.D.Disturbed in Penobscot Bay's prehistoric Native American village

It is now one the most visited factors of interest at the Maine State Museum. It isn't clean if Norsemen made it that a long manner.

There had been no distinct Norse artifacts located in Maine. Perforation on the coin shows that it have become worn as an ornament and made its manner from Newfoundland thru coastal commerce.

New England might also moreover have been domestic to first-rate Old World civilizations.

The Romans seized the Phoenician settlements of what is now Morocco and had many precise ports along the Atlantic coasts of North Africa. They additionally managed the coast of Portugal. Was it viable for an adventurous Roman to look out over the massive ocean and wonder what grow to be beyond?

New England has visible scattered Roman artifacts, but there are even though questions about provenance and authenticity. Although some Roman coins have been determined inside the United States (including Maine and Texas), they were now not excavated at an archeological net page.

Because ancient cash are a fave of lenders for centuries they'll have been brought to New World via Colonial or Modern people, and then lost.

Some might have been delivered to New England with the aid of the usage of manner of Roman ships that had been blown out of path and ended up over the sea. These coins are plenty awesome from the Native American artifacts, and could were taken into consideration curiosities and change gadgets, which encompass the Maine Penny.

The proof for pre-Columbian colonization of New England is not sturdy, as you could see. Archaeologists even though expect the smoking gun, clearly as they did with Native Americans migrating to New England in advance than the Bering Strait Land Bridge.

Chapter 9: Aliens and UFOs

The Phoenicians and Celts aren't all extended-distance voyaging those who may additionally have visited New England. Others could have been from further afield, which include from outer place.

Maine is the extremely good country to have had a UFO sighting. Cynthia Everett have end up a Camden teacher who saved a magazine inside the route of the primary years of the nineteenth Century.

Her writing regularly describes the every day sports sports inside the northernmost country, such as college sports and harvest. Although it's miles exciting reading for historians, it has nothing to be awe inspiring till you attain the July 22 1808 get admission to:

"Around 10 o'clock, I noticed a totally uncommon appearance. It modified right into a colourful mild that got here from the East. It regarded to be a Meteor at the begin, however I fast determined out that it turned into now not.

It seemed to transport as fast as light and look like within the surroundings. However, it dwindled within the path of the floor and persevered to adventure at the same distance, every so often ascending or descending. It then moved round inside the Horizon (it wasn't very moderate), and then it lower back to its authentic function.

Could this be a herbal phenomenon or a person-made one? Everett had no clarification and have become knowledgeable. She observed the night time time time sky every night time, in evaluation to most humans these days. She did not see the streetlights and pollution.

She modified into very familiar with the planets and stars, and did no longer make commonplace mistakes like many present day UFO spotters. Many cutting-edge UFO critiques communicate of lighting fixtures moving spherical, changing colours at the horizon, and then disappearing.

It modified into discovered that the horizon is constantly the western horizon and they have been simply seeing a glittery celebrity, or planet, diffusing through Earth's environment, warmth radiating from the floor. Due to the placing of the celestial object, the "UFO" disappears. I attempted to discredit a sighting of this nature thru the usage of a firsttime camper however to no avail.

Everett turned into an professional observer who must document herbal phenomena collectively with earthquakes or the appearance of comets. Her son became captain of a clipper vessel and may have used the celebs to navigate. It could no longer sound like a meteor, celebrity, or planet. There are not any reports within the media that advocate she sought hobby for her sighting. This could have been an early UFO sighting. Ufologists receive as true with so. Her description of the sighting has been considerably stated as the first documented UFO sighting in the United States. But it might not be the very last.

Since then, fantastic gadgets have been visible in our skies on many activities, mainly inside the center to late twentieth centuries. On the night of September 19-20 1961, New Hampshire changed into the scene of one of the most first-rate instances in the united states of america of america.

Barney and Betty Hill drove along New Hampshire's Seacoast Highway that night time time time as they decrease again to

Portsmouth from a adventure to Niagara Falls. The avenue changed into almost deserted at 10:30 p.M. Betty located a awesome moderate passing beneath the moon and Jupiter as they drove through Lancaster on U.S. Route three. It rose, then started to transport erratically. Intriguing, she requested her husband for assist. Barney pulled the car over at Twin Mountain's picnic region and the couple were given out.

Barney and Betty Hill

Betty took out a pair binoculars, and set them up at the moderate. She changed into bowled over to look a disc-fashioned item with multicolored lighting.

It modified her existence.

The couple had been stunned and had been given lower back of their car. They drove slowly to check the ordinary craft that appeared to be following them. Betty observed that the craft have become at maximum one-and-a 1/2 of times the peak of the herbal granite rock formation referred to as The Old Man of the Mountain. This technique it become at least 60 toes in period.

They drove on and the UFO followed them until they reached Indian Head. At that factor, the craft dropped and hovered over them for approximately a mile.

Barney were given out of the auto and stopped it. Barney used his binoculars to appearance that the UFO had an open window via which he should view numerous human beings looking lower again at him.

These figures appeared more humanoid and wore matching caps and glossy black uniforms. All however one in all them moved faraway from Barney's window. He regarded to be speaking into his thoughts and telling him to "stay there and hold looking." A lengthy form jutted out from the bottom UFO.

Barney panicked at that issue. Barney ran once more to his vehicle shouting to his partner, "The creatures plan to capture them!" Barney stepped on the gasoline and the pair sped down.

Betty glanced up, but couldn't see the UFO. She furthermore couldn't see stars, regardless of it being a clean night time. It changed into almost as if there was something blocking her view.

The vehicle vibrated, sending a tingling sensation through every of them. The couple moreover heard everyday buzzes, beeps and specific sounds.

They fell asleep and have been woken up via a few exceptional collection buzzes and beeps. The couple had vague reminiscences of factors of the road,

146

turning onto a facet street and being stopped by way of a fiery item. They had traveled approximately 35 miles, unbeknownst to their companion.

They all over again home at sunrise, many hours after their anticipated arrival, with none reminiscence of the time misplaced or stopping for an prolonged time period.

They went to mattress exhausted, but determined a few ordinary information the following day.

Barney had scuffed his shoes and the strap of his binoculars modified into snapped. These subjects were now not his fault. Betty's robe have become torn and guarded in unidentifiable red powder. She, too, had no clarification.

They located a sequence concentric circles of exceptional, concentric circles inside the trunk once they checked out the auto. The needle spun after they exceeded a compass close to the circles, as although it modified into being suffering from a sturdy magnetic place.

Betty cautioned the incident to the community Air Force Base. Officials from the Air Force took word of her document and positioned it in Project Blue Book. Project Blue Book is the Air Force's valid database for UFO sightings. Betty furthermore wrote to Donald Keyhoe, a famous Ufologist approximately her evaluations.

The couple later visited a hypnotherapist to get better their out of area memories. Benjamin Simon, a hypnotherapist, interviewed every couple separately over numerous taped education. Both felt traumatized and afraid at a few level within the commands. Simon had to prevent the consultation several times to calm them down.

Barney's hypnosis observed to him that he felt an splendid urge to tug off the road as he drove far from the alien thru the window. Six figures surrounded the auto while he pulled off.

Under hypnosis, he modified into terrified through the creature's eyes and said, "Oh, those eyes." They are in my brain. They are actually there. They are simply up to my eyes, urgent in the course of mine."

They took them aboard the deliver, and separated them into unique rooms. Barney was then placed on an exam desk. The terrified man saved his eyes near, but they might sense his backbone, genitals and anus. They additionally took a pattern of his pores and skin.

Betty had a similar enjoy however stored her eyes open. Both of them heard aliens talk in a language they couldn't recognize however it sounded nearly like they have been mumbling. Barney heard them speak English into his head, on the equal time as Betty's examiners spoke to Betty verbally, even though without fluency.

A movie star map end up showed to her that confirmed wherein the aliens originated. Some stars have been connected with the useful resource of using strains that marked interstellar exchange routes.

Simon believed that the Hills had been mendacity and that the hypnotist did no longer consider in aliens.

The Hills to start with did no longer are trying to find media exposure but they fast decided it. The Boston Traveler reporter observed out approximately the hypnosis classes, and managed to get Simon's notes. The October 25, 1965 tale have end up an worldwide media sensation. This case have become possibly to have obtained more interest because of the truth that the Hills were a blended race couple (Barney Hill grow to be African-American whilst Betty Hill become white and that they every participated in the NAACP), that may be a extraordinary phenomenon in the United States.

The Hills have been now unfastened to tell their whole tale.

John Fuller, a UFO researcher, helped them to put in writing The Interrupted Journey. It has been a bestseller in UFO literature due to the fact that 1966.

It changed into so famous that it attracted the attention of Astronomy mag. They published an alien celebrity chart and requested readers to analyze the sorts of stars to determine which part of the universe it represented. It might also show the celebrities close to Zeta Reticuli and the "trade routes" that converge on these stars.

Barney Hill exceeded away in 1969. Betty Hill survived till 2004. For a few years, she decrease decrease back to the kidnapping internet web page on-line numerous instances in keeping with week to search for UFOs. Although she claimed to have seen them fly through the sky frequently, they by no means picked her up over again.

The activities of that night time time time were preserved in the UFO Canon. They have turn out to be so famous that the New Hampshire u.S.A. Erected a historic marker at wherein the 2 New Englanders had been said to had been kidnapped.

Chapter 10: Strange Creatures

New England is domestic to many splendid and unusual critters, further to extraterrestrial beings. The Lake Champlain Monster is the most famous, a resident of Lake Champlain and located among Vermont, New York.

Champy, moreover referred to as Champy, has been sighted over 3 hundred times. Some sightings date again to the Native American length. Legends of a weird creature in the lake have been shared through the usage of both the Iroquois in addition to the Abenaki.

It modified into referred to as "Tatoskok" with the resource of the use of the Abenaki. Champy is a miniature version of his famous cousin from Loch Ness. He has an extended neck and a small head. The head and neck are typically the simplest visible components above water.

Benjamin Radford's instance of Champy

A description of the monster is given in a 1883 account. Sheriff Nathan H. Mooney claimed that he saw a "substantial serpent of water" approximately 50 yards from him.

The serpent changed into amongst 25-30 ft long and the sheriff determined first rate white spots inside its mouth. P.T., the circus owner, located this early sighting and started to get maintain of quite some evaluations.

Barnum provided $50,000 for Champy, but nobody ever observed him. Barnum needed to provide the monster to everybody else, so he ended up having different exhibitions.

There have been many sightings on account that then, further to three mysterious pics and films. Cryptozoologists are human beings who have a study unknown animals and declare that it can be a surviving Plesiosaur. These marine dinosaurs have been concept to had been extinct at some stage in the extraordinary extinction sixty five million years in the past. Many cryptozoologists receive as proper with that they will be although living in Loch Ness and Lake Champlain.

A Discovery Channel excursion hired Fauna Communications Research to document sound at 3 locations at the lake in 2003. These recordings recorded what sounded just like echolocation.

This is a valid utilized by marine mammals collectively with porpoises and whales to speak and discover their manner. Echolocation isn't a commonplace feature in Lake Champlain.

The sightings flow on for a hundred twenty 5 miles. The maximum remote sightings of the Lake Champlain Monster are my very personal, heaps of miles a ways from his home. I modified into getting to know a e-book approximately Ethiopia as soon as I got here inside the direction of a Champy shirtwearing nearby man. I changed into greatly surprised to say the least. I modified into then capable of see one-of-a-type American t-shirts that promoted traveler locations and village gala's.

One even got here from my alma mater the University of Arizona. Clothing companies buy t-shirts that are not in use from the us and promote them to Ethiopians at rock bottom charges.

Some charities acquire garb to donate to the poorest humans on this colorful, however in spite of the fact that horrific u.S.. Champy, an first-rate pal and colleague in Addis Ababa remains remembered.

Champy is not the most effective New England lake monster. Willy, a 23-foot-lengthy serpent from Vermont's Lake Willoughby, may be placed at Lake Willoughby. Willy grow to be captured, not like the Champy. According to the August 14, 1868 trouble of The Caledonian-Record, Stephen Edmonds, Newport, VT, a twelveyear-vintage boy, killed Willy the outstanding water snake at Willoughby Lake. He ran boldly over to the monster and severed its body using a sickle. The actual measurements decided out that the quantities measured 23 toes prolonged.

 Edmonds' seize have become not accountable for Willy's loss of lifestyles; sightings hold to at the existing time.

The most famous sea serpent in America lurks in reality off the coast Massachusetts. It is regularly visible near Gloucester, a fishing and shipping port.

The serpent has been noticed in the region for the purpose that 1638 and once more as recently as 1997. It is described as a serpent-like creature measuring as plenty as eighty ft in period, with a head the dimensions of a horse protruding of the water vertically at the pinnacle of its body.

John Josselyn, in 1638, modified into the first to give an explanation for this beast. He stated he heard about it but never located it. Josselyn said, "They advised u.S. Sea serpent [sic] that lay quoiled up like a cable on the rock at Cape Ann.

A boat handed through with English aboard and two Indians.

They have to have shot him, but the Indians dissuaded the crew, pronouncing that if he have been not killed, all is probably in hazard."

It became observed on numerous events inside the seventeenth and18th centuries.

There had been even more sightings in 1817. Many of those opinions are from sailors, in comparison to many sea monsters that are visible with the useful resource of landlubbers.

Solomon Allen III, Shipmaster, testified that his head seemed like a rattlesnake's, however modified into nearly as big as a horse's. His motion on the water floor have become sluggish. He would possibly occasionally bypass in circles and special times at once earlier.

Two days later, each other deliver noticed it. Matthew Gaffney, a deliver's wood worker, tried to seize the creature.

He was probably wondering that he can be a hero the numerous women or make a few coins selling tickets.

Maybe he grow to be terrified of sharing the water and with this type of huge animal. According to him, he said, "I had a exceptional gun and I took accurate purpose." I shot at his head and I don't forget I hit him. After I fired, he have become toward us and I concept that he modified into coming at me.

But he went immediately beneath our boat and made his front at approximately a hundred yards from the spot wherein he had sunk.

All evaluations have the same smooth abilities: the vertical feature and collection of humps in water in the again of the pinnacle that propose a snake-like frame. Over the subsequent many years sightings have been common, but they bogged down inside the 20th century.

Since 1997, there has now not been a sighting. The final one modified into in 1962. Are the sea serpents lengthy past or have they died? The disappearance of sea serpents will be because of the truth that the region isn't able to keep the local monster populace.

Although the Gloucester Sea Serpent, and exclusive watery creatures, will be inclined to be left by myself, it isn't always a pleasing feeling to percentage your swim with an unknown creature from the depths.

Many human beings may want to as a substitute see them lengthy gone, or within the approach of demise. The globsters might be an example of this.

Even despite the fact that many human beings have seen them, and some of them were photographed, no character sincerely is aware about what globsters appear like. Because pleasant lifeless and rotting globesters had been ever discovered. Many of the useless our bodies were washed up on seashores, however some fishermen have controlled to catch the pungent creatures of their nets.

Globsters regularly look like blobs or flesh with the tentacles on the ends and lengthy necks.

Scientists receive as authentic with they will be each rotting whales, or octopi. DNA attempting out discovered that the Nantucket Blob have become the remains of a whale. It modified into determined off the coast of Nantucket Island in Massachusetts.

Basking sharks are each distinctive possible globster, with a bigger mouth and a thinner, decrease jaw. Basking sharks rot, and the lowest of their mouths fall away, leaving in the lower back of a creature similar to an prolonged-necked sea monster.

Vermont seems to have an relatively immoderate extensive fashion of monsters. The Pigman of Northfield is one of the maximum sinister. A nearby farm boy vanished in 1971.

He become in no manner placed. People notion he had fled or been abducted at first. But, soon, they started out out to suspect a few issue more sinister.

They observed a person absolutely naked, included in white fur, and numerous other sightings. He have emerge as approximately five'10", and his face regarded like a pig.

The sightings were focused spherical an abandoned pig farm close to Union Brook and a thick, dank woodland known as "The Devil's Washbowl," wherein there were many herbal caves that locals tended no longer to visit. Many drivers referred to seeing a white guy taking walks thru the woods at night time time as they drove through the Washbowl. One driving pressure claimed that the creature leapt onto his car's hood and glued its pig nose up in competition to the windshield.

Local investigators appeared across the location and located what seemed to be crude beds made from branches and leaves at the Devil's Washbowl and at an antique pig farm. Although the investigators braved the vicinity, they did not find the creature. He have become greater lively at night time time.

People started out to keep away from the Devil's Washbowl and some teenager couples started out out using it as a Lovers' Lane.

This turn out to be a horrific concept for one couple who've been gambling of their automobile at the same time as the boy needed to relieve himself of a far much less pressing need. The Pigman attacked him as he were given out of his car to look for a tree. The creature attacked him with its claws and then threw him in the direction of his car collectively together with his cloven hooves. After that, the Devil's Washbowl emerge as a great deal much less famous as a parking spot.

It seems that the Pigman's best 2nd became inside the course of this assault. Soon after, sightings began out to lower. He hasn't been seen for many years. Is he vain or hiding someplace?

The many evil creatures that stay on Vermont's ski slopes border on the absurd are almost comical. Mogul Monsters and Snow Snakes lurk inside the powder, equipped to journey unlucky skiers. These mischievous monsters have an extended data. Blaming them for your incapacity to balance is better than blaming you!

Monsters aren't truely found in Vermont. All over the Northeastern United States, strange creatures are commonplace. Cassie, a sea serpent from Maine, has been sighted off the coast of Maine because of the fact that 1779. The most ultra-modern sighting happened in 2002. It is believed to be a serpent with a length of forty five-100 feet. Many fishermen and sailors have seen it, who're familiar with sea creatures and can no longer mistake something else for a sea serpent.

The Dover Demon, a terrifying creature discovered in Massachusetts, is any other. Bill Bartlett, a teenager, and of his buddies first saw this creature in 1977 at the identical time as driving to town at night time. Bartlett modified into using even as he determined the bizarre creature trekking up a nearby stone wall. It become three to 4 toes tall and walked on all fours. However, it regarded humanoid. Each limb had 4 extended, tapering fingers and become prolonged and skinny.

Bartlett claimed that the pores and pores and skin come to be peach-colored. However, it is not smooth how he might also want to tell on the same time as driving by using manner of in terrible lighting fixtures. He defined the pinnacle as being melonshaped and about the equal size as his frame due to the truth the maximum bizarre function.

 The creature grow to be now not seen by way of the usage of the usage of the lads who had been inside the automobile with them.

Someone else, however, did. John Baxter, a close-by youngster, furthermore noticed the same creature that night time time. It emerge as no longer a protracted manner from the primary sighting. It come to be additionally spotted thru Abby Brabham and Will Taintor the subsequent night time time time.

Paranormal investigators have determined a similarity among Dover Demons and alien Grays, which is probably regularly noted in UFO opinions. Folklorists say it resembles a Native American creature referred to as Mannegishi. The description seems like a child moose to Naysayers, however what fun might that be? Not to say the reality that Massachusetts isn't always recounted for its wild population of moose. It sounds to me like a chupacabra (the well-known "goat sucker"), that is a not unusual sight in Texas and the Desert Southwest. There stays no motive of what Dover's young adults found in 1977 on those dark nights.

Another Lovers' Lane legend speaks of the ugly Melon Heads in Connecticut. These are a few bizarre-searching humanoids which have massive heads. They are stated to be high-quality three feet tall in step with a few reports, however others claim they may be everyday top. They had been accused of committing terrible crimes in competition to unwary patients who wander into their territory. Many Connecticut towns have seen Melon Heads, at the aspect of Milford and Shelton, Monroe. Seymour. Weston, Oxford. Southbury. Trumbull. These roads also are referred to as Melon Head Roads via nearby teenagers.

Saw Mill City Road, wherein Melon Heads out of doors Shelton had been visible

Most humans receive as proper with they will be human but inbred. They have stored their secrets for the purpose that Revolutionary War. It isn't clean why they've remained so isolated from society. Some declare they're mutants, escaped intellectual patients or inhuman monsters. There are many horrifying memories (regularly unprintable), approximately what they'll do to you in case you get caught on their turf.

Many of those tales involve teens from all over the u.S.A. Of the us. Most of them use the Lovers' Lane as a records. Back inside the day of the strain-in, some traditional B-movies featured monsters attacking couples parked in vehicles. These instances have a big quantity of male witnesses. These recollections may had been told to distract their more youthful siblings. This isn't always something that would discourage adults from reading. Stories like the ones first rate enchantment to mother and father who are curious and worried. These memories may additionally additionally had been meant to discourage the women from getting out the auto, or possibly the eyewitnesses really observed some factor because of their misperception or excessive weed or alcohol consumption.

We should not disregard those tales too rapid without doing a little more research.

Chapter 11: Hauntings

Ghosts aren't viable in settlements as ancient as the ones found in New England. It is domestic to many ghostly stories from those who've returned to the opposite aspect, and it is one of the maximum lovable regions inside the u . S ..

The Old State House in Hartford is Connecticut's oldest capital. It offers a captivating tour and is designed in Federal fashion. The historical façade hides an inexplicable spirit. The zero.33 floor was opened to Joseph Steward, a community painter, who's seemed for his pictures highlighting the maximum essential figures of the time. In 1797, he opened the primary museum in the u . S . A ., a group of curiosities. His collection protected sideshow factors of interest including a - headed porc, a filled alligator and a unicorn horn. Museums had been a hodgepodge affair decrease returned then. These museums had been a way to carry the area to people within the days earlier than the net and tv. Although the gathering modified into subsequently dismantled, it changed into reassembled within the Nineteen Nineties with new devices. It is now a museum for the curious.

Old State House in Hartford

Steward, because it seems, come to be no longer able to detail along along with his studio and museum. His footsteps can be heard inside the route of the constructing and his shadowy silhouette is frequently noticed among the well-knownshows.

The nation government moved to the Victorian Gothic-fashion country capitol in 1878. It is thought to be haunted. According to legend, William Buckingham, the preceding governor, lurks in Room 324. Security guards are regularly awoken with the beneficial resource of the sound of footsteps within the night time time time and discover that there can be no individual inside the building. Strangely, Buckingham died three years preceding to the final contact of the constructing's manufacturing. It is feasible that he draws to the area because of the statue of Buckingham near the western front.

Slater Mill, Pawtucket in Rhode Island is each different ancient building that has a haunting presence. It have become built in 1793 and is called the birthplace the American Industrial Revolution. It is now a National Historic Landmark and may be toured to reveal the creativity of an technology past.

Twelve-hour shifts have been labored on the manufacturing thru youngsters as younger as six years vintage

Floor, with simplest a half-hour harm. Because of their smaller, greater agile fingers, toddler exertions became favored because of the fact they were better prepared to govern the thread spinning on the quick-shifting machines. It moreover meant many children misplaced their fingers or their lives to injuries as there had been not safety recommendations decrease lower again then.

These youngsters appear to be the deliver of many ghosts that hang-out Slater Mill. One of the ghosts is a small lady who may be seen looking for at the lawn from the building. A barefoot boy has additionally been seen walking up and across the corners, sprinting out of sight. Both children are carrying oldfashioned garb. Staff members at Slater Mill are open about hauntings and characteristic many hair-raising testimonies to percentage. One guy, whose honor it's miles to shut down the area at night, modified into walking up the wood stairs sporting heavy art work boots. He heard a girl say, "Quiet." Although he turned into on my own on the time, he did what he grow to be informed, and removed the boots. Visitors to the excursion record listening to voices and feeling touched via invisible hands even within the course of sunlight hours. As a give up end result, Slater Mill is one of the maximum haunted homes New England has ever seen.

However, now not all ghosts stay interior. Maiden's Cliff, Maine is the hold-out of a bit female who tragically died there. Eleanora French, an eleven-yr vintage woman, climbed Mt. Megunticook accrued her buddies to enjoy the lovable view of the lake, and surrounding geographical region. It became a lovable spring day. The best day to move on an excursion.

Megunticook, sporting red, raced ahead of her buddies to select out wildflowers.

Then, a wind blow up. Eleanora's head became blown off through manner of the wind. She raced to capture her hat and ran over the three hundred-foot cliff!

Eleanora did not die immediately, however she turn out to be in horrible pain even as she have emerge as taken once more to her circle of relatives. Later that day, she died. Today, hikers and boaters at the lake see a bit female sporting pink choosing wildflowers up the slopes. Her red hat is floating within the lake under.

Chapter 12: The Witch Who Got Away

It is widely believed that the early colonial duration become dominated thru strict Puritan religion and a narrow view on morality. This also can have been authentic for some groups, but many New England farmers lived in relative isolation away from any church. Many did no longer observe prepared religion and lived their lives consistent with their private guidelines. Modern ministers bitch that many families did no longer have a Bible at their homes and are not interested by attending divine services. It changed into difficult to enforce spiritual orthodoxy because of the small amount of people scattered all through this sort of big rural location. The wave of witch trials that swept over colonies emerge as now not handiest an assertion of Church strength, but additionally a quit stop end result of superstitious worry about the black arts.

Salem, Massachusetts have grow to be the maximum well-known case, which came about in 1692. However, there had been many witch trials at some stage in New England starting in 1642 in Connecticut. The colony moreover passed a law making witchcraft capital offense in that identical yr. Exodus 22.18 states, "Thou shalt in no way allow a witch stay." Connecticut finished four witches in 1662. Most of the humans accused in the American colonies have been women. However, a few guys had been tried and determined responsible.

An example of the Salem Witch Trials

Many human beings view the Salem Witch Trials as a unmarried event, in which some human beings from one village allowed their fear to get the higher of them, and sentence harmless people to loss of life. Understanding the Puritan enjoy of the seventeenth century is important at the manner to recognize why, how, and even as the Salem Witch Trials came about.

The Puritans arrived in Massachusetts in 1620 thinking they had been going to the Promised Land. They believed they had been promised it. They decided a rugged coast that grow to be surrounded through using the Atlantic Ocean for optimum of the 12 months. They were one crop a long manner from starvation due to the reality the growing season end up shorter than in England. Their kids could in all likelihood all die earlier than conducting adulthood. Those who survived have been often problem to highbrow and bodily pains because of a healthy dietweight-reduction plan collectively with warm sermons and bloodless food.

Two villages had been installed within the middle of the 17th century, now not a long manner from the famed Plymouth Rock. Salem Town changed into the primary and protected 600 humans, lots of them from Puritan households. Salem Village come to be Puritan as well, however more than half of of of its six hundred populace have been associated with the Putnam family. It separated from Salem Town in 1684, at the request the Putnams.

Many different households from the Village had been now not happy with the choice and it delivered approximately tensions every spiritual and secular. Salem Village and Salem Town had an extended information of disputes over property strains. Early ministers in Salem Town complained approximately the metropolis's disability to pay them. Other issues came on the equal time as Samuel Parris become chosen as Salem Village's first ordained minister.

It turned into clean that the real covenanters' laws were not enough to save you each shape of sin. Salem Village delivered additional prison tips and rules to cover the entirety, from the manner a female wears her hair to the strategies someone makes a living. The Puritan fathers had to choose among admitting defeat or blaming a person else on the same time as their children did no longer comply with the ones criminal recommendations. The first opportunity modified into impossible for a holy people, so the second needed to be the purpose. America became a place wherein the satan emerge as at big, similar to England. Their nice preference of salvation have emerge as to discover his fanatics and deliver them once more to their evil draw close.

Everyday human beings located out from religious leaders on the ones subjects, specifically in Puritan Massachussets, wherein secular political leaders had been essentially a theocracy. Puritans prohibited music, dancing, or joyous celebrations of holidays. Their kids had been taught only faith. Cotton Mather, a Boston-based totally minister, posted Memorable Providences regarding Witchcrafts & Possessions (1689), in which he defined how witchcraft had affected some of Boston youngsters. His beliefs and works have been in a function ignite a firestorm.

As it's far now, rooting out evil changed into a whole lot easier than it emerge as over again then. Witch trials were as not unusual in England at some level within the English Civil War as homicide trials. Although some of the sufferers have been a great deal much less lucky, the death with the resource of setting modified into the equal punishment. The conflict changed into over and Oliver Cromwell was made Lord Protector. However, the massive form of accused started out out to drop. In 1685 Charles II changed Charles II collectively with his brother James II. This caused accusations falling dramatically as Puritans who remained in England were pushed underground with the useful resource of persecution. Ironically, because of the fact the witch trials in England have been waning, they have been only simply beginning in America.

Importantly, a few bodily symptoms and signs and symptoms that human beings were experiencing within the seventeenth century had no rationalization. The Puritans couldn't provide an explanation for why a person all at once felt pain in their backs or how they were able to yell out at will. So, buddies accused pals. For centuries, historians debated over the motives of the accusations. They were both to oppress ladies or gain economic benefit.

www.ingramcontent.com/pod-product-compliance
Lightning Source LLC
Chambersburg PA
CBHW060221030426
42335CB00014B/1302